Voices of All Day DevOps, Volume 1

FEEDBACK LOOPS

ISBN: 9781796989724
Imprint: All Day DevOps Press

Publisher:
All Day DevOps Press
48 Wall Street, 5th Floor
New York, NY 10005

www.alldaydevops.com

Voices of All Day DevOps, Volume 1

FEEDBACK LOOPS

Praise for All Day DevOps: The World's Largest DevOps Conference

"30,000 registrants. Wow!"

"New nuggets of ideas emerged, good conversation was had, and it was great to put faces to names for those of us who have conversed virtually but had not yet met in person."

"It's superb that such an excellent program is shared worldwide and free of charge to anyone interested."

"I learned so many new things and I've met amazing people. I'll be forever grateful. Looking forward to next year."

"Please hurry and bring it back again, I can't wait to attend!"

"Too much good information and sessions at once! I found myself wishing I could watch them all. I am so glad you make the recordings available!"

"Loved it — I need to have the rest of my team get involved earlier next year."

"Keep this level of awesomeness!"

Table of Contents

INTRODUCTION .... **IX**
by Derek Weeks

CHAPTER 1: A DevSecOps Field of Dreams..1
presented by Julie Tsai

CHAPTER 2: Ann Winblad Reflects: The Rise of Software.................5
presented by Ann Winblad

CHAPTER 3: Automating Chaos in the Pipeline 9
presented by DJ Schleen

CHAPTER 4: Sometimes You Just Need a Pencil and Paper 15
presented by Boyd Hemphill

CHAPTER 5: Cancer Sucks. DevOps Helps.. 19
presented by Sarah Elkins

CHAPTER 6: Characterizing and
Contrasting Container Orchestrators... 23
presented by Lee Calcote

CHAPTER 7: Continuous Control with Continuous Delivery............ 31
presented by Sherry Chang and Edward Harris

CHAPTER 8: Continuous Everyone: Engaging
People Across the Deployment Pipeline 37
presented by Jayne Groll

CHAPTER 9: Creating an Appsec Pipeline in a Week...................... 43
presented by Jeroen Willemsen

CHAPTER 10: DevOps and Security is Like Smoking Meat 49
presented by Apollo Clark

CHAPTER 11: DevOps for Normals 53
presented by Michael Coté

CHAPTER 12: DevOps for Small Organizations:
Lessons from Ed 57
presented by Ed Ruiz

CHAPTER 13: DevOps: Building Better Pipelines 61
presented by Dan Barker

CHAPTER 14: DevOps: Escape the Blame Game 65
presented by Matthew Boeckman

CHAPTER 15: DevOps: Making Life on Earth Fantastic 71
presented by Helen Beal

CHAPTER 16: DevOps: Making the Boring Things Stay Boring..... 75
presented by Mykel Alvis

CHAPTER 17: DevSecOps: Overcoming
the Culture of Nos with Chaos 79
presented by DJ Schleen

CHAPTER 18: Docker Image Security for DevSecOps..................... 85
presented by José Manuel Ortega

CHAPTER 19: Docker: The New Ordinary 89
presented by Daniël van Gils

CHAPTER 20: Tickets Make Ops Unnecessarily
Miserable: The Journey to Self Service.. 93
presented by Damon Edwards

CHAPTER 21: From "Water-Scrum-Fall" to DevSecOps...................101
presented by Hasan Yasar

CHAPTER 22: How Capital One Automates Automation Tools 105
presented by George Parris

CHAPTER 23: How To Fully Automate CI/CD — Even Secrets 109
presented by Andrey Utis

CHAPTER 24: Increasing the Dependability of DevOps Processes 113
presented by Ingo Weber

CHAPTER 25: Alert, Alert: Alert, Alert: Monitoring is More Than You Think 117
presented by Jason Hand

CHAPTER 26: Leading a DevOps Team at a Fortune 100 Company 121
presented by Uldis Karlovs-Karlovskis

CHAPTER 27: Microcosm: Your Gateway to a Secure DevOps Pipeline as Code 127
presented by Hasan Yasar

CHAPTER 28: Modern Infrastructure Automation 131
presented by Nathen Harvey

CHAPTER 29: Nothing Static About the Growth of Static Analysis 135
presented by Justin Collins

CHAPTER 30: One Team, 7,000 Jobs: Life in the DevOps Jungle 141
presented by Damien Coraboeuf

CHAPTER 31: Organically DevOps: Building Quality and Security into the Software Supply Chain at Liberty Mutual 145
presented by Eddie Webb

CHAPTER 32: Scaling DevOps at Pearson 149
presented by Sean D. Mack

CHAPTER 33: Securing Immutable Servers in a Serverless World 153
presented by Erlend Oftedal

CHAPTER 34: System Hardening with Ansible 157
presented by Akash Mahajan

CHAPTER 35: The DevOps Trifecta 161
presented by Sumit Agarwal

CHAPTER 36: The DevSecOps Equilibrium 165
presented by Chris Corriere

CHAPTER 37: The Road to Continuous Deployment 169
presented by Michiel Rook

CHAPTER 38: ABN AMRO Embraced CI/CD to Accelerate Innovation and Improve Security 175
presented by Stefan Simenon

Introduction

Three years ago, I was walking the halls with Mark Miller at a DevOps conference in San Francisco and the energy of the community was palpable. About six hundred people were in attendance, sharing their knowledge, exchanging ideas and making connections. Some of us were years deep into our DevOps journey and others were just beginning — but everyone, regardless of experience, was welcome.

You see, one of the things I have grown to love throughout my high-tech career has been the learning experience. I remember very early on in my career, sitting at an industry conference where Scott McNealy, then head of Sun Microsystems was keynoting. "You know what is so remarkable about this industry," he remarked, "to be good, you can never stop learning." At the time, his comment raised my enthusiasm for the career I had chosen to pursue, and as the years has passed, I've reflected on the enthusiasm that statement raised in me many times.

In the twelve months preceding that walk down the hall, Mark and I had probably been to 25 DevOps conferences. Like in San Francisco, the other conferences introduced us to so many of industry thought leaders, advocates and practitioners. We were seeing the same people over and over, extending our conversations, and sharing lessons learned since our last encounters. But for all of those people we had met, we recognized that they were often the only person from their organization that we interacted with. Travel budgets, time away from the office, and seniority barriers kept hundreds — sometimes thousands — of their colleagues away from the experiences we were all sharing.

There had to be a better way.

The content, speakers, and hallway conversations at these conferences were not built to educate everyone. Scale was limited. The organizational impact was shallow. Education was offered to a select few.

This sparked an idea. I remembered a few years prior when Mark had championed a "Follow the Sun" conference for the Microsoft SharePoint community. Mark was lining up speakers from all over the world to share their experiences in an online forum. At the time he had done that, SharePoint conferences and meetups were happening all over the world on what seemed like almost a daily basis. You could feel the communities' energy from all directions. With the momentum and energy I felt surrounding the DevOps community that day, I thought we could produce something similar to that "Follow the Sun" concept.

I turned to Mark and said, "What do you think about bringing this entire experience that we're having here in San Francisco to an online forum, where everyone could participate?" His eyes widened and without a second of hesitation he said, "Let's do it."

This was the moment All Day DevOps was born. The wheels started turning.

Some things explicit from the beginning. We wanted to provide access to everyone — so the conference needed to be online. More importantly, it needed to be free. Lack of budget or barriers of geography could not stand in the way of knowledge.

We had also experienced all too many conference presentations where someone was giving a (sometimes not so) veiled vendor pitch. We never liked those and the audiences we were a part of felt the same way. All Day DevOps would be vendor-pitch-free from day one.

While the concept of All Day DevOps was born in the halls of San Francisco, I remember the day Mark came back to me with a reflection on scale. While other online conferences had been attempted from a couple of software vendors here and there, they were constrained to about four hours of online interactions filled mainly with vendor-biased content. Mark operated on a different realm and was the one who laid out our initial schedule of 15 hours, over 15 time zones, and an agenda that would play home to 57 speakers. Nothing of that scale had ever been attempted, even back in the SharePoint community.

To pull this off, we invited a few of our friends from the community who we had met at DevOps conferences along the way. Shannon Lietz (Intuit), Andi Mann (Splunk), Karthik Gaewad (Oracle), Ernest Muller (Alien Vault), Chris Corriere (Autotrader), James Wickett (Signal Sciences) and Milton Smith (Oracle) joined us to formulate the conference program, execution and speaker roster. If all went successfully, we told them that we could pull off the biggest DevOps conference ever, with over 1,000 people participating.

Four weeks after we announced the concept of the conference in our Call for Papers promotion, we knew we were onto something. With barely more than a meetup widget for registrations on the homepage of AllDayDevOps.com and the slogan "57 speakers, 15 hours and 15 timezones", we had over 300 people register. Little did we know that 60 days later, we would have over 13,000 people participate on the day All Day DevOps went live on-air. That day was November 15, 2016.

All Day DevOps has offered countless hours of learning opportunities since that day. Not only did we deliver those 57 sessions, but we opened up a digital hallway track with the help of our Slack channel. I believe we recorded over 25,000 conversations on our first day there.

The very next year, we expanded All Day DevOps to the world. We decided to run for 24 hours, offering 100 free practitioner led sessions, accompanied by the hallway track on Slack. In 2018, the community born concept of viewing parties in local communities had grown to over 135 locations around the globe while over 30,000 participated online.

This book is a compilation of the stories gathered from All Day DevOps sessions. These are the stories shared by our community of practitioners that helped me and countless others learn more about DevOps. The challenges, the frustrations, the opportunities, and accomplishments were captured. The nuances, tricks of the trade, and secrets from the front lines were all shared to help us learn from the journeys of others. We met the enlightened who had made it to the promised land and heard from the pioneers with arrows in their backs who had not had the smoothest of journeys.

This is the good stuff. The things Mark and I were learning ourselves at those in person events, that we wanted to bring to everyone, line these pages.

As with all journeys, there is never a single path to success. The journey through this book is not meant to be sequential, but one where hops, skips, and jumps are part of the fun. Pick the stories that best apply to the experiences you want to learn from and dive in. Learn from your peers, chat with your colleagues about the stories you read here, and reflect upon how they might improve your own journey.

Perhaps your All Day DevOps experience started with us in 2016 or maybe it is just launching today. Whatever the case, we invite you to learn with us and share with others. One day, if you are so inclined, you might even want to join us as a speaker for All Day DevOps. Everyone, *yes everyone*, is welcome.

Derek Weeks
Co-Founder, All Day DevOps

CHAPTER 1

A DevSecOps Field of Dreams

presented by Julie Tsai

CHAPTER 1
A DevSecOps Field of Dreams

While millions of people love baseball, the same can't be said for security and compliance — well, at least not yet. Perhaps one day.

Much like in the immortal baseball movie, Field of Dreams, if you build a friendly security and compliance system, the developers and operators will come. At least, that is the contention of Julie Tsai (@446688), Head of Infosec at Roblox.

Good Architecture Sustains Sound Applications and Security

Julie's contention is that you can build a system that might actually bring a little joy to developers and operators, and it starts with realizing that, at the end of the day, we are all looking for good architecture. Good architecture sustains sound applications and security. It makes everyone's life easier — so we all have a little time for baseball (or football or board games or even curling).

The good news is that DevOps organizations are ahead of the game. Julie pointed out the old joke, "DevOps, isn't that just where you give the keys to the developers?" Well, no, and it isn't a specific tool or deployment. For security, it is about being lean on requirements, building compliance in from the start, and integrating it across the development lifecycle so you can build something that is secure and performant. Besides being best practices for developers, you can also use your code and policies to appease your auditors as compliance is built in. Bottom line — DevOps puts the rigor around security and compliance.

It Takes a DevSecOps Village

Julie points out a couple of takeaways from DevOps that help security and compliance:

- You can own a problem and be an individual leader. You can think globally but act locally by seeing across the silos and bringing a message of empowerment.
- It provides a path towards integration to internalize other groups' values, bring them into your own words and ways, and mutually reinforce and thrive. It also informs how people work together.

Remove Friction to Scale DevSecOps

The goal here is to ultimately get us something easier to scale and maintain and be compliant and secure. Julie outlines steps to get you closer to this idyllic system:

- Use configuration management because it leads to precision and more unified and verifiable work.
- Get rid of whatever you can to eliminate mistake factors. You will also get to a place to streamline workflow so there are fewer mistakes.
- Extend infrastructure as code. It can become configuration as policy and your audit trail can connect the intent to the execution.

DevSecOps: Where Things Go Wrong

With these in place, Julie contends you need to realize the world isn't perfect, so you need to build a system that injects security and compliance where it is most effective. To find the areas in the development lifecycle that it is important to inject, ask these questions:

- Where is someone intending to make a change?
- When you are releasing it live into your production stack.
- What is happening when something isn't going as expected? Is it being changed back to where you need it to be, are you monitoring it, or are you doing nothing?

DevSecOps: String Together Wins

You also need to make it simple. She quoted Mark Burgess, "IT has a detail sickness," noting that, "We are often burdened by complexity — we love it and dig right in, but it is important to understand the level of granularity you need. You need to look for the things that can make a critical uplift that gives you an incremental improvement. String together the wins."

In the end, Julie says to, "try to reach a goal of being visible and streamlined and leverage the automation and technology in a way that is joyful in how we use it. It is not about a rigid process that is going to die soon — it is about what we are trying to bring to the whole world so that we work more efficiently and more technical."

CHAPTER 2

Ann Winblad Reflects: The Rise of Software

presented by Ann Winblad

CHAPTER 2
Ann Winblad Reflects: The Rise of Software

Ann Winblad started her own software business when most people didn't know what software was. It was 1976, and she borrowed $500 from her brother. Six years later, she sold her company for $15M — a year before the first Mac was released. Since 1989, long before investing in software was trendy, she co-founded and serves as the managing director for Hummer Winblad Venture Partners. In the years since, 81 of their investments have been acquired or gone public. We are also proud to have her sit on our Board at Sonatype.

As successful software venture capitalist, Ann "auditions the future" everyday, and she talked about the waves of digital disruptions and what it will take to succeed tomorrow.

Ann shared that when she started investing in software development, she was told that it was too risky. We have all lived through the rest of the story. Ten years ago, one out of the top 10 highest valued companies, Microsoft, was a tech company. Social media concepts were just emerging. The term DevOps was only coined 10 years ago by Patrick Debois and Andrew Shafer (August 2008). Today, 7 of the 10 most valuable public companies are tech companies.

Technology is leading the charge on a global scale and as Ann observed, "data is the new oil."

It is these companies that are investing in the future. It is these companies that are driving a massive wave of innovation, spinoffs, and new businesses, and massive digital disruption. Imagine this: the 5 U.S. tech companies are annually investing $60 billion in R&D — close to the non-defense R&D budget of U.S. Government.

What Ann really focused on, though, is the fact that software development — for any company — must become a core strategic competency, not just a cost center, or they risk being Ubered or Amazoned.

Ann continued by citing the Jeff Bezos concept that Amazon will always be a "day one" company. That is, one that always acts like it is on its first day — always hustling, always focused. Of course, Amazon sets a high bar, being highly automated with continuous software development. They average 136,000 software deployments per day and push code to production every few seconds. Contrast that with data from a Forrester survey Ann cited — only 34% have complete automated product lifecycles and less than 20% release faster than monthly.

To make software development part of your enterprise's core competencies requires a strong focus on automation, agility, quality, security, reuse, and speed in the software supply chain. The good news is that DevOps practices embrace all of these concepts.

Ann contends that DevOps is key to making software a core competency.

This is more than moving from Waterfall to Agile. Ann notes that the jump from Agile to DevOps is not as obvious as you think, "Some assume that Agile is all about processes while DevOps is about technical processes. This points you in the wrong direction by placing them in separate streams in the transformation. DevOps breaks down silos, it does not create new ones. DevOps strives to focus on the overall service of software delivered to the customer, and it breaks down barriers between software and operations teams."

What does Ann think is up next? A category 5 hurricane for the enterprise in cloud, mobile, AI, and big data.

Are you ready?

Ann said that she started her career was a coder, and was not considered a "developer." But she reflected that developers are now in a primary position to drive businesses forward, saying that "developers today are business strategists."

Are you a strategist?

No matter your business, Ann contends that software is at its heart, if not its soul. She advises that the rise with software is the win in the near market. This isn't about increasing the velocity of software development. It is about understanding your company's core competencies, knowing how you generate value, and continuously codifying the opportunity ahead. You also need to increase customer engagement and operational efficiency.

Ann closed with, "My job is to audition the future. Your job is to create the future. Are you fighting the trends or are you inventing the trends?"

CHAPTER 3

Automating Chaos in the Pipeline

presented by DJ Schleen

CHAPTER 3
Automating Chaos in the Pipeline

You implement security in an unobtrusive way and increase the quality of the product when it hits the customer.

Sound too good to be true? Well, it isn't, and DJ Schleen (@djschleen) lays out how to do it. DJ is a DevSecOps Advocate at Sonatype and a former Security Architect and DevSecOps Evangelist with Aetna.

DJ begins his talk by making the point that DevOps allows us to disrupt the traditional mindset of security — the culture of no. Rather, we can deeply integrate security with dev and ops so that security becomes part of everything we do on a daily basis. He describes this as, "an unprecedented opportunity" because it:

- Disrupts traditional security approaches
- Defends against modern attack vectors with innovation
- Fosters creativity, collaboration, and culture
- Adopts automation and tools
- Instills repeatable processes, resiliency, and scalability
- Facilitates continual feedback and easily auditable processes
- Allows anyone to commit to production as fast as possible
- Demands application as code, infrastructure as code, and security as code
- Reduces production operational impacts
- Reacts quickly to software vulnerabilities and their remediation

Goals

With the benefits laid out, DJ talks about the goals of placing security into the pipeline. He states, "It is more than just automating the scan button. It is about ensuring software passes through a well-defined and automated set of gateways that assess code security without decreasing velocity but can stop the pipeline when critical vulnerabilities are detected and manual intervention is necessary." It is also about getting actionable remediation guidance back to those writing the code, sharing as much information as you can, and not forgetting about code after it hits production.

Vital to achieving all of these goals, though, getting to a culture of yes and a culture of let's work together.

People

We all know that all of this isn't possible without people. DJ reminds us of the formula, success = people + process + tools. People are the one who create processes and select the tool sets. Invest in the right people.

Of course, many people inherently fear change. You have to be cognizant that some people are cautious, others are progressive. How

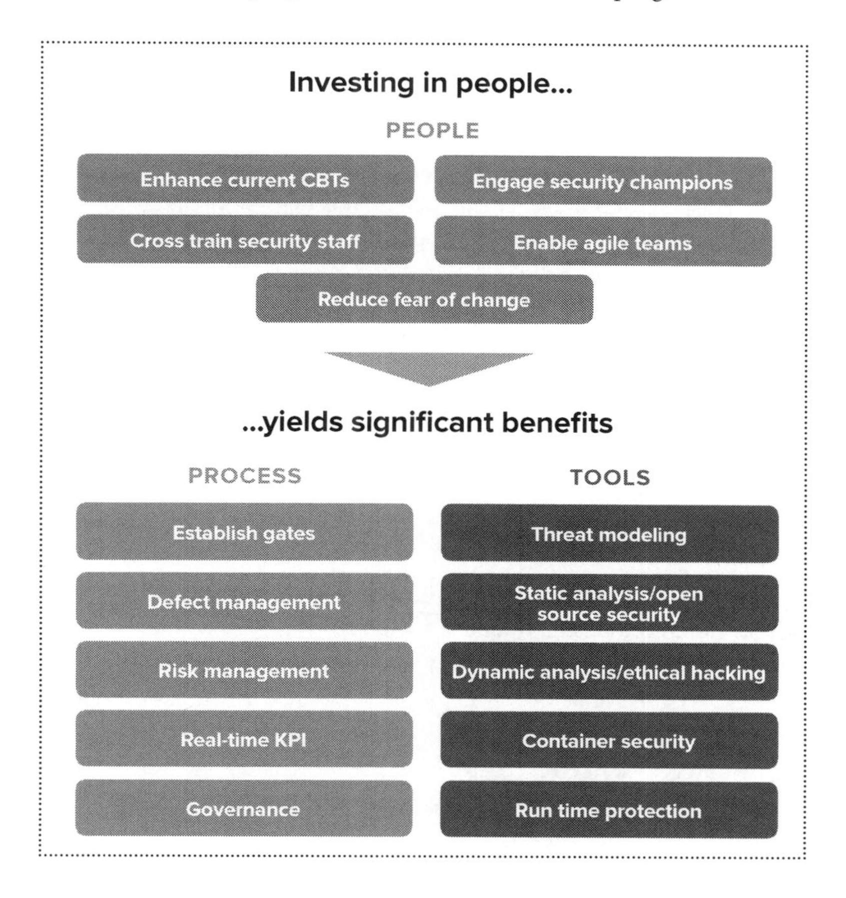

can you implement security automation with the least resistance to change? Ask yourself where you can put security gates, such as static code analysis or dynamic analysis tools, without disrupting the flow or the velocity of your teams.

Is Agile Agile Enough?

DJ asks the important question, "Is Agile agile enough?" Can it adapt quickly enough to address security concerns or is there a better way? DJ's concern is that many Agile organizations go from waterfall to mini-waterfalls, while DevOps, "breaks us from the chains of waterfall." Since DevOps allows you to continuously push out changes, you can respond much quicker. You don't have to wait 2 weeks for the sprint to end. This makes it so much easier and effective to react to security vulnerabilities.

Evaluation

You wouldn't drive your car blind, and you shouldn't drive your pipeline blind. It is a wealth of knowledge, and you need to set and then gather information on key performance indicators (KPIs). Use your security requirement gathering tools to define your security requirements before ideas are developed. They can integrate with testing penetration tools to ensure the security requirement has been met when the code enters the pipeline.

Third-Party Tools

Look at the tools you are using because third-party tools mean you are bringing in code that not just you know, but everyone in the world knows. With containers and virtual servers, remove what you don't need. Do you need a bluetooth driver if you are running a web server?

Introduce Chaos

We still need the normal checks, but we need to introduce chaos. You should randomly take containers down and exercise production environments because small environments don't accurately simulate large systems. As DJ said, "Do it in production. Live large." You can use tools like Kube-Chaos and Pumba to test resilience by randomly killing containers. While it requires a mature application and infrastructure, a resilient system will have no client-facing impact and will make it difficult for an attacker to maintain a foothold.

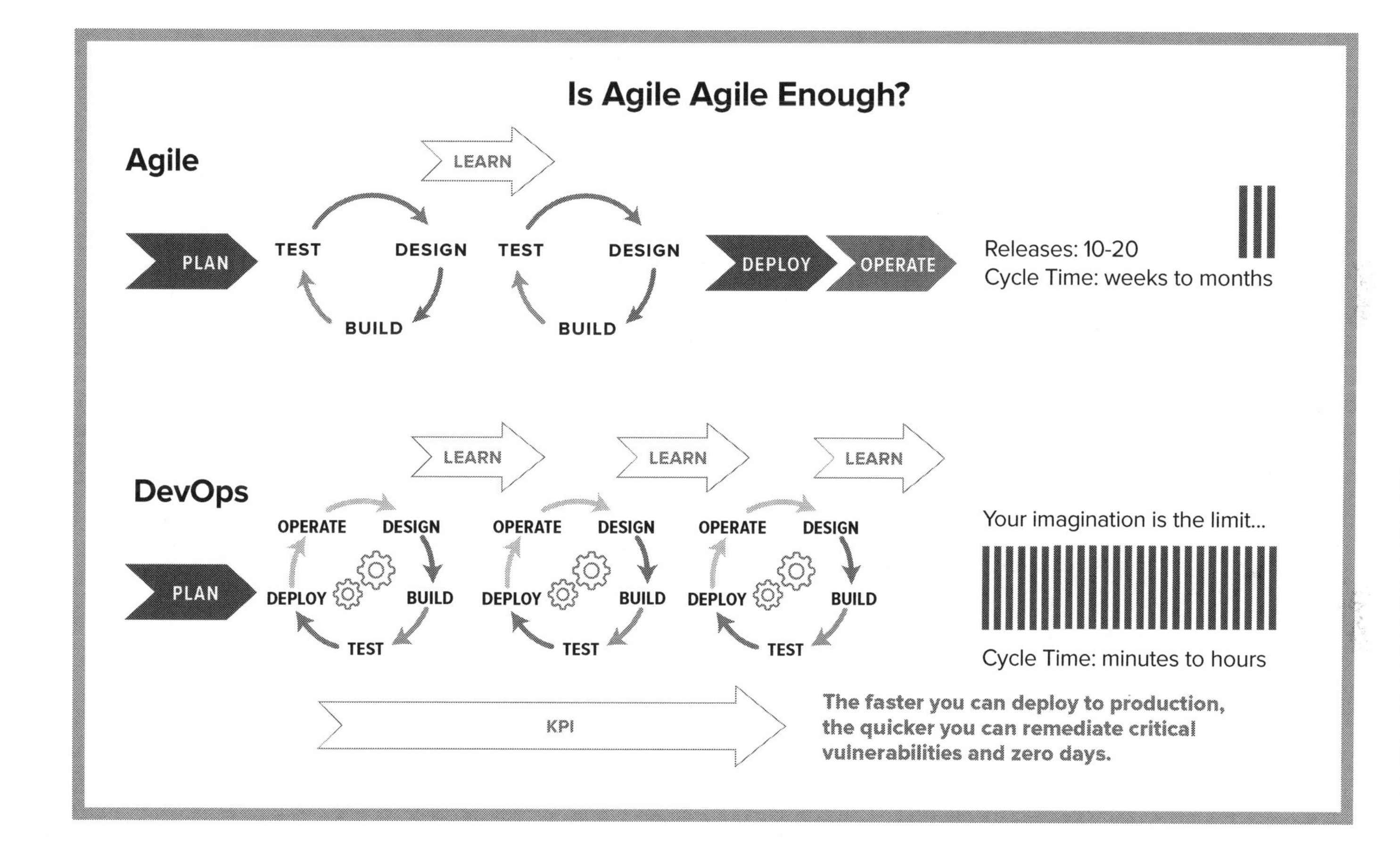
Is Agile Agile Enough?
Agile
LEARN
PLAN
TEST
DESIGN
BUILD
TEST
DESIGN
BUILD
DEPLOY
OPERATE
Releases: 10-20
Cycle Time: weeks to months
DevOps
LEARN
LEARN
LEARN
PLAN
OPERATE
DESIGN
DEPLOY
BUILD
TEST
OPERATE
DESIGN
DEPLOY
BUILD
TEST
OPERATE
DESIGN
DEPLOY
BUILD
TEST
Your imagination is the limit...
Cycle Time: minutes to hours
KPI
The faster you can deploy to production, the quicker you can remediate critical vulnerabilities and zero days.

Shuffle the Deck

DJ underscores that if you are using anything open source, everyone knows what is there. Everything becomes repeatable, so it is easy to distribute an exploit. To thwart this, every time a container starts, scramble the image to create a moving target.

It's Not Easy — But it's Possible

Of course, every good thing has challenges. DJ outlines some:

- Integration of security tools in a manner that supports objectives for actionable continuous feedback
- Choosing the toolsets that provide the least disruption as possible to DevOps teams
- Extracting the proper KPIs and indicators from the process and making them actionable
- Tailoring the people, processes, and tooling to unique environments
- Teaching old dogs new tricks
- Multiple "flavors" of DevOps in different parts of the organization
- Introduction of chaos requires application and infrastructure maturity
- Traditional security folks don't code

Takeaways

- Don't fear deploying rapidly and often into production
- Always gather information in the form of KPIs and make them actionable
- Support the organization with tools, techniques, and best practices
- Automate. Everything.
- Defects are defects — regardless if they are a code defect or a security vulnerability
- Code your infrastructure — eliminate access to physical or cloud based machines
- Choose tools that interfere minimally with flow
- Introduce chaos to become a moving target

As we can see from DJ's experience, security automation is an investment with challenges, but one worth making.

CHAPTER 4

Sometimes You Just Need a Pencil and Paper

presented by Boyd Hemphill

CHAPTER 4
Sometimes You Just Need a Pencil and Paper

I once heard a time management instructor maintain that most time management systems fail because you are trying to force a type A system onto a type B person. Sure, sophisticated, color-coded, multi-tool systems work great for some people, but others do best with a piece of paper and a pencil.

The same holds true for software development. Boyd Hemphill, former CTO of Victory CTO and current Director of Cloud Operations at Contrast Security, walked through examples from his professional career when the latest and greatest tools and processes were not the best solution.

Boyd started with Feedmagnet, a social media aggregator. Back in 2011, they had one MySQL database for all of their clients. Imagine the load on that database when one customer had lots of chatter on Twitter or Facebook. It meant one customer success caused many customer failures. So, they needed to create something that was "anti-fragile." They signed a large customer for SXSW and decided to clone what worked. It worked, so they started cloning the system for each customer rather than put everyone into the same bucket.

> "Fools ignore complexity. Pragmatists suffer it. Some can avoid it. Geniuses remove it."
>
> — ALAN PERLOS

> "The business schools reward difficult, complex behavior more than simple behavior, but simple behavior is more effective."
>
> — **WARREN BUFFET**

At the W2O Group they were building custom WordPress microsites. It was 2014. Boyd explained how they we were always late and it was always way too expensive because their system was far too complex. This meant customer changes were too expensive and took too long — all business problems. This is something DevOps recognizes — that, first and foremost, it is a business problem.

For the W2O Group customers, single tenant was a given. So, they standardized development environments, enabling Continuous Delivery from day one. They used Vagrant and Chef to standardize WordPress environments in production so developers could switch between projects in 5 minutes rather than 8 hours, which is what it used to take them. Devs requested changes via Chef code and pull requests. Because this is a pattern they are familiar with, it made it easier on them and adoption less painless.

Furthermore, having standard development and production environments coupled with a standard way of working fostered effective communications between dev and ops — you know, the whole goal of DevOps. Business engagement skyrocketed and they went from losing customers to being given repeat work. The recovered 8 hours per week in lost work time translated to $1.8M for a $60M business, and they recognized that customers cannot host sites themselves, so they charged a premium for hosting sites, generating $1.2M in new revenue.

His main takeaway is, "DevOps — the new often makes the old more possible." When you are serving businesses and not consumers, consider single tenancy, relational databases, and technology you can hire easily.

"It does look similar—but this one is powered by Hadoop."

CHAPTER 5

Cancer Sucks. DevOps Helps.

presented by Sarah Elkins

CHAPTER 5
Cancer Sucks. DevOps Helps.

Cancer sucks. But with folks like Sarah, DevOps is helping make a difference in the race to a cure.

Sarah Elkins is not curing cancer herself, but she is employing DevOps practices to help those who are. Sarah supports the technology infrastructure for those who are trying to cure cancer at the National Institute of Health (NIH).

Sarah Elkins (@configures) configures technology solutions at the National Cancer Institute (NCI), where she has worked for 9 years. NCI is a federal agency — part of the NIH. They support over 700 websites, from basic HTML to content management systems to complex bioinformatics systems which have multiple tiers and thousands of servers. They use multiple operating systems, containers, and physical and virtual machines (both on-premise and some AWS). Developers range from lone scientists to large teams.

The NCI is automating its builds and deployments — showing it is possible even in a large bureaucracy, and they use a variety of processes and technologies to enable software to move from source code repositories all the way to production servers. This includes GitHub, Jenkins, Nexus, and more, with a variety of teams involved.

We Need to Talk

Sarah's cure begins with **engagement** — development, infrastructure, and security all working together — just as teams of specialists work with patients to help them beat cancer.

To help speed development and approvals, a necessary evil that can grind development to a halt if not managed well, they agree that applications drawing on an existing technology catalog are approved as operational and security teams provide guidance to development on what is required for an Authority to Operate (ATO). In other words, they are trying to pre-approve as much as possible and communicate earlier in the process.

DevOps Practices are Maturing

Source code is primarily kept on GitHub, including a public repository for non-proprietary source code, and they primarily use Maven or Apache Ant for **build scripts.** Infrastructure teams provide XML templates to provide consistency.

Most NCI software relies on **build dependencies** on open source software components. They use artifacts during builds and utilize Sonatype's Nexus as their repository manager.

All of this supports the **automation of builds and deployments** at the NCI. For builds, most active projects are either on, or migrating to, Jenkins. Build artifacts may be .zip, .war/.ear, or Docker images.

For **application deployments**, they use development, quality assurance, stage, and production stages. Most teams use some form of automation, from simple (copy content, stop/start container) to more complex scripts. They allow developers to perform deployments for robust applications, and some manual orchestration is required, for instance for database timing or related applications.

As an organization, they are moving towards Continuous Integration, with varying progress among teams.

Building Security In

The have also embraced involving **security** early, often, and automatically. They have deployed self-serve/on-demand Nessus scans, which allow developers to see, on their own, how they are doing as soon as the application is stable. Security teams run AppScan and Twistlock for Docker image scanning during Jenkins builds, just before deploying, and frequently scanning the repository in between. For issues found, development and infrastructure teams work together to remediate security concerns.

At the end of her talk, Sarah offered up three takeaways:

1. DevOps at NCI is a work in progress (isn't it everywhere?)
2. The wide-ranging needs at NCI require flexibility, communication, and teamwork
3. There is no shortage of work

Since it is a work in progress, what opportunities are ahead at NCI?

- More automation for individual applications with Jenkins
  - » Developers can perform container restarts on lower tiers
  - » Database updates (some projects are using Liquibase already)
  - » Some applications are still manual/batch scripted
- More Docker and improving Jenkins / Docker instances
- More orchestration automation and Puppet/Ansible integration
- Security improvements through
  - » Scan dependency artifacts before building
  - » More integration and automation

Are you in a similar organization with a looming bureaucracy and/or regulatory environment standing the in the way of DevOps. Be encouraged, make a note of what has worked of them, and keep moving forward.

CHAPTER 6

Characterizing and Contrasting Container Orchestrators

presented by Lee Calcote

CHAPTER 6
Characterizing and Contrasting Container Orchestrators

Admiral Calcote — also known as Lee Calcote (@lcalcote) or the Ginger Geek to his friends — knows containers and does containers and talks containers.

Okay, he isn't really an admiral — nor does anyone call him that —but he used the title admiral to describe what container orchestrators do, relating it to an admiral directing a fleet of container ships. You could also say that they are like the conductor of an orchestra, directing the individuals to work together as a group toward a common goal while each musician is still able to play their own instrument.

Lee Calcote is the Founder of Layer5, and for his talk, he walked through four open-source container orchestrators: Nomad 0.5, Swarm 1.12, Kubernetes 1.5, and Mesos 1.1 .

Core Capabilities

- **Cluster Management**
  - Host Discovery
  - Host Health Monitoring
- **Scheduling**
- Orchestrator Updates and Host
- Maintenance
- Service Discovery
- Networking and Load-Balancing
- Multi-tenant, multi-region

Additional Key Capabilities

- Application Health Monitoring
- Application Deployments
- Application Performance Monitoring

He emphasized the obvious — there is no one perfect solution. Each organization is different, so for each solution, he looked at:

- Genesis and purpose
- Support and momentum
- Host and service discovery
- Scheduling
- Modularity and extensibility
- Updates and maintenance
- Health monitoring
- Networking and load balancing
- Secrets management
- High availability and scale

Lee noted that while there are many core capabilities, any orchestrator must have cluster management and scheduling.

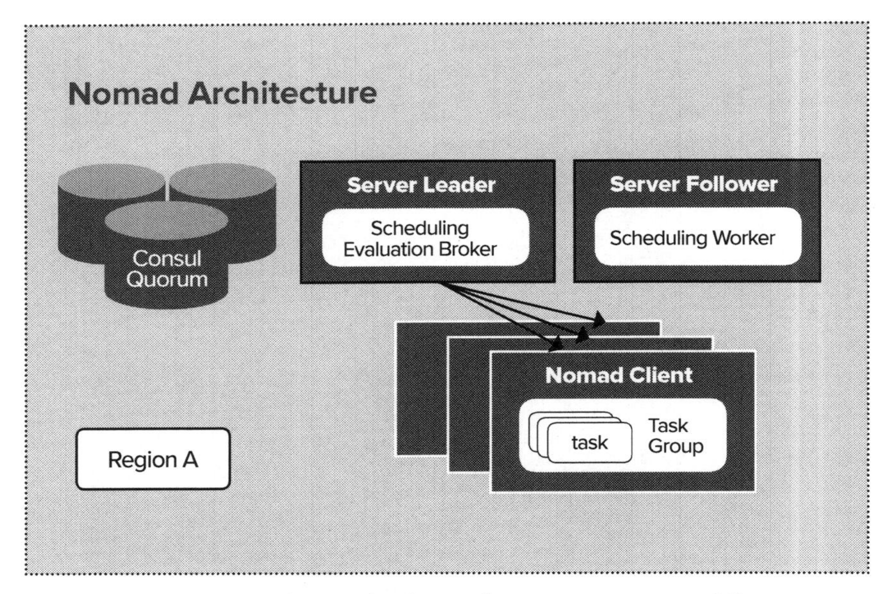

He then dove deeper into the four solutions. Summaries follow:

Nomad

- Designed for both long-lived and short-lived batch processing workloads
- Cluster manager with declarative job specifications
- Ensures constraints are satisfied and resource utilization is optimized by efficient task packing
- Supports all major OSs and workloads
- Written in Go and with a Unix philosophy
- Host discovery: Gossip protocol — Serf is used; servers advertise full set of Nomad servers to clients; creating federated clusters is simple
- Service discovery: Integrates with Consul
- Scheduling: two distinct phases — feasibility checking and ranking; optimistically concurrent; three scheduler types when creating jobs
- Uses task drivers to execute a task and provide resource isolation, but it does not support pluggable task drivers
- Built for managing multiple clusters/cluster federation

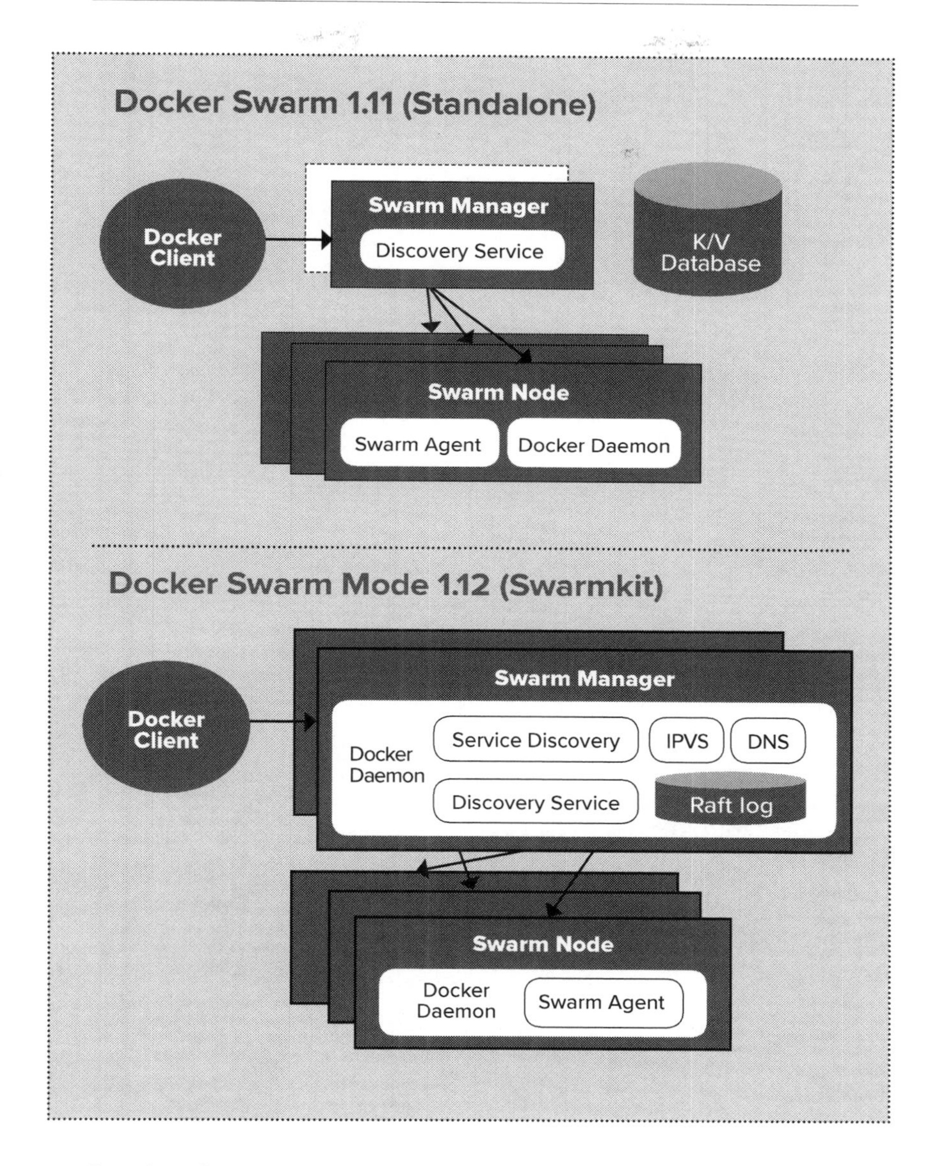

Docker Swarm 1.12

- Simple and easy to setup
- Architecture is not as complex as Kubernetes and Mesos
- Written in Go — lightweight, modular, and extensible
- Strong community support

- Host discovery: used in the formation of clusters by the Manager to discover Nodes (hosts); pull model — worker checks-in with the Manager
- Service discovery: Embedded DNS and round robin load-balancing
- Scheduler is pluggable and is a combination of strategies and filters/constraints
- Ability to remove "batteries"
- Rolling updates are supported
- Managers may be deployed in a highly-available configuration, but does not support multiple failure isolation regions or federation

Kubernetes

- An opinionated framework for building distributed systems
- Written in Go and is lightweight, modular, and extensible
- Led by Google, Red Hat, and others

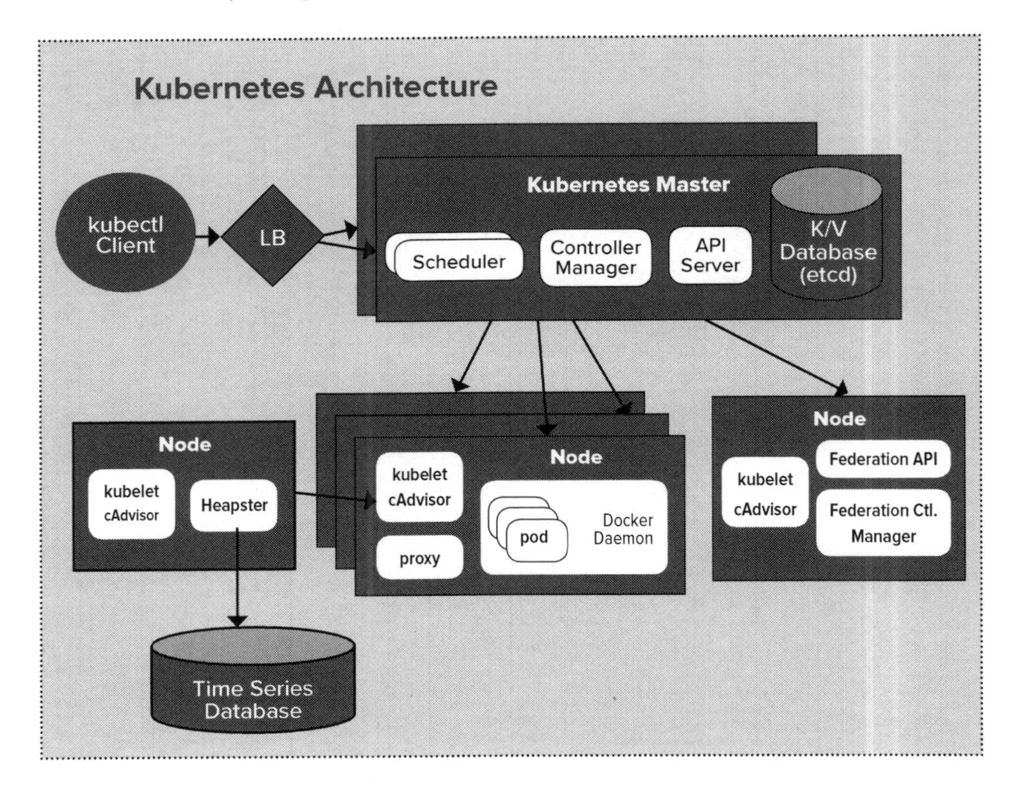

- Young — about two-years-old
- Robust documentation and community
- Scheduling is handled by kube-scheduler
- Pluggable architecture and an extensible platform
- Choice of: database for service discovery or network driver and container runtime
- Supports rolling back deployments, automating deployments and rolling updating applications
- Inherent load balancing
- Uses Pods, an atomic unit of scheduling. Each pod has its own IP address, no NAT required, and intra-pod communication via localhost

Mesos-Marathon

- Mesos is a distributed systems kernel
- Mesos has been around the longest (since 2009)
- Mesos is written in C++
- Marathon is a framework that runs on top of Mesos
- Mesos is used by Twitter, AirBnB, eBay, Apple, Cisco, and Yodle
- Marathon is used by Verizon and Samsung
- Mesos-DNS generates an SRV record for each Mesos task
- Marathon ensures that all dynamically assigned ports are unique

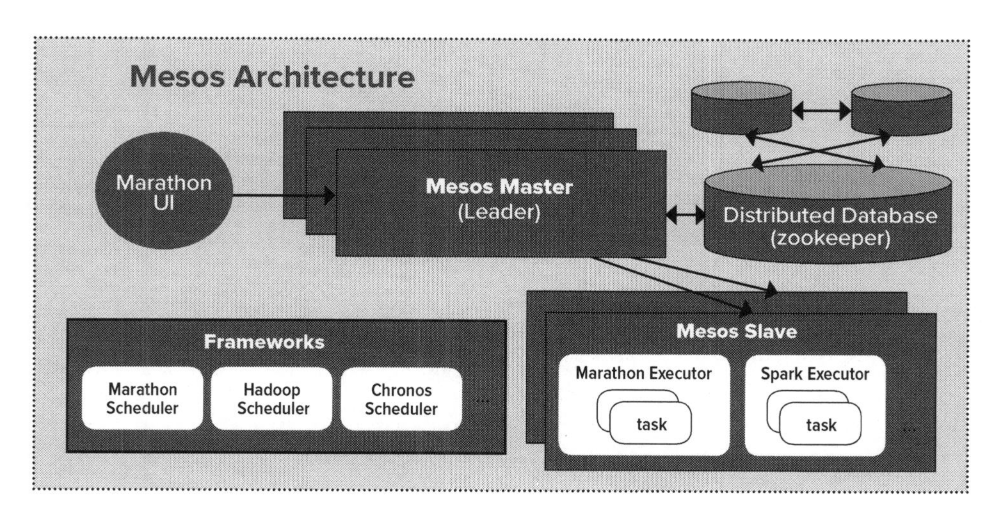

In the end, Lee provided the following overview comparing the different container orchestration solutions — well, at least trying to compare apples and oranges.

A High-Level Perspective of the Container Orchestrator Spectrum

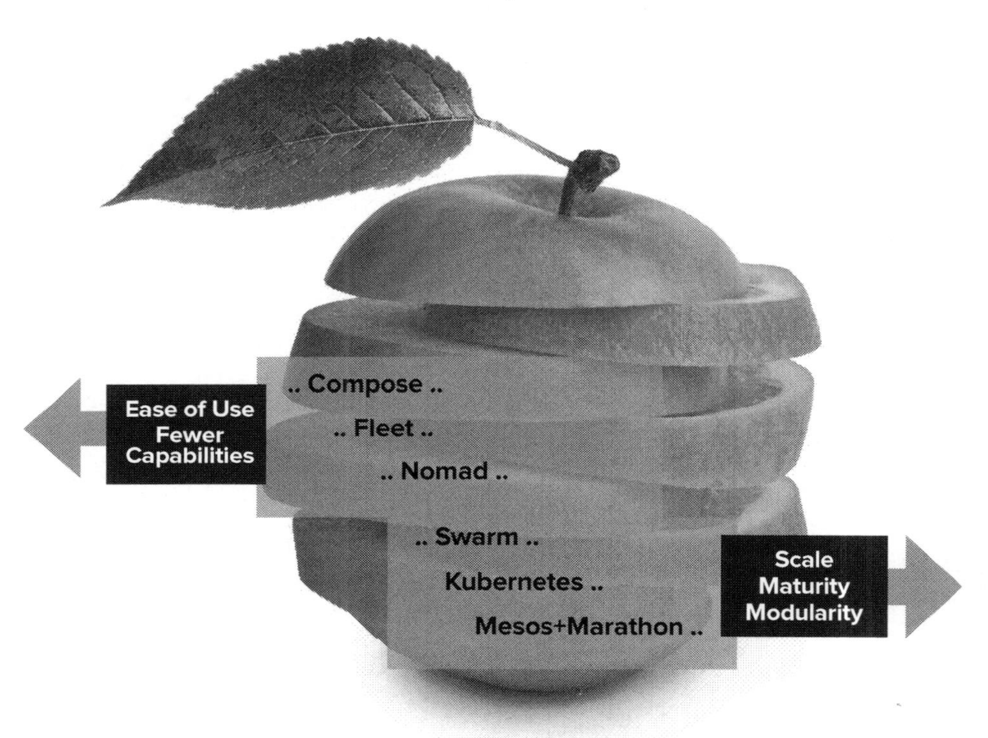

CHAPTER 7

Continuous Control with Continuous Delivery

presented by Sherry Chang and Edward Harris

CHAPTER 7
Continuous Control with Continuous Delivery

Anyone working within a software development organization that works with change control board (CCB) has been frustrated by the bureaucracy. Perhaps they think innovation grinds to a halt because of the CCB, or hours are burned with bureaucracy. They might long to work free of a CCB. Whatever varied opinions people have of CCBs, they are rarely credited as instruments of innovation.

Thus, the question, "With Continuous Integration/Continuous Delivery (CI/CD), can you get rid of the change control board?" This was the question raised by Sherry Chang (@sherrychangca) and Edward Harris (@EdwardHarrisJr).

Sherry and Edward both have decades of experience in IT and work at Intel. Given the size of Intel, they have a CCB, yet they were successful in integrating CD at Intel. This is the story of how they did achieved agility and speed with compliance, governance, and decision gates.

Now some of you may be reading this and thinking, "everyone is 100% on board with CI/CD at our organization." (You can stop reading and begin preparing your 2018 All Day DevOps conference presentation — because you may live in rare air that others could learn from). For the rest of you living with entrenched processes and people, Sherry and Edward's story should help you break the log jam and reorient the river.

Sherry and Edward begin by restating some inherent truths about the incompatibility of compliance within continuous delivery. In their traditional form, these are two disciplines at odds with each other:

First, Compliance Demands

- Documentation requirements
- Manual review of changes and signed offs
- Feedback cycles that depend on meeting cadence

CD Demands

- Software deployable at all times
- Push-button deployment anytime
- Fast, automated feedback on production readiness of systems

As Sherry noted, on small projects, no one really cared. But when it came to mission-critical projects, you run into walls.

How do you breakthrough the wall? Speak a common language and paint a clear picture of what is good and what is bad. If you can add in change tracking, audit capabilities, and testing, it will create an opportunity to automate. It is just centralizing the data they are already gathering.

You also need to look at why we need approvals for change, and how automation can improve compliance. The below chart shows the reasons for change approvals and the benefits of automation for each one. The bottom line is that automation improves compliance, and information security review backlogs can be reduced or eliminated by automating some of the reviews.

Now, teams at Intel can automate goals and just review exceptions rather than the standard. That changes compliance from a necessary gate that delays delivery to more of an enabler to meet quality and security goals.

How Can You Get Started?

Option A: Hybrid of Manual and Automated Releases. First, map the processes and then work out a roadmap to automate some of the tasks. Everyone needs to agree where to invest by figuring out where the greatest pain point (bottleneck) is. This process gives you invaluable visibility into the process.

Pros: Clear ownership of each step; evolve towards full automation.

Cons: One size fits all, agnostic to team maturity.

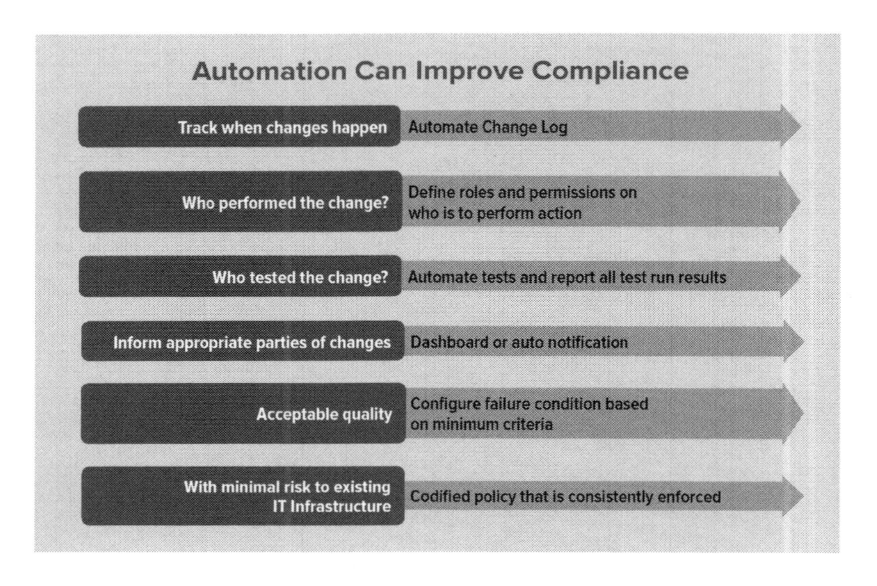

Automated Quality Gate

Policy Monitor	Qualification Criteria	Automated Quality Gate`	Audit`
Software Quality	• Code coverage % • Cyclomatic Complexity Index • Coding Standards and Convention • Defect density	• Test Execution Result • Static Code Analysis • Code Coverage	• Test Run Report • Code Coverage Report/Trend • Code Review of Automated Tests
Security	• No critical or high priority issue discovered	• Security Vulnerability Scanner	• Scan Result
Configuration Compliance	• No blacklist technology (i.e. Java Client, Silverlight)	• IP Scanner	• Scan Result • Puppet Manifest
Product Heath	• Minimum Acceptable Incidence and Outage • Minimum Acceptable Performance	• # of incidents • # of outages	• Production Monitoring Report • Incidents Volume

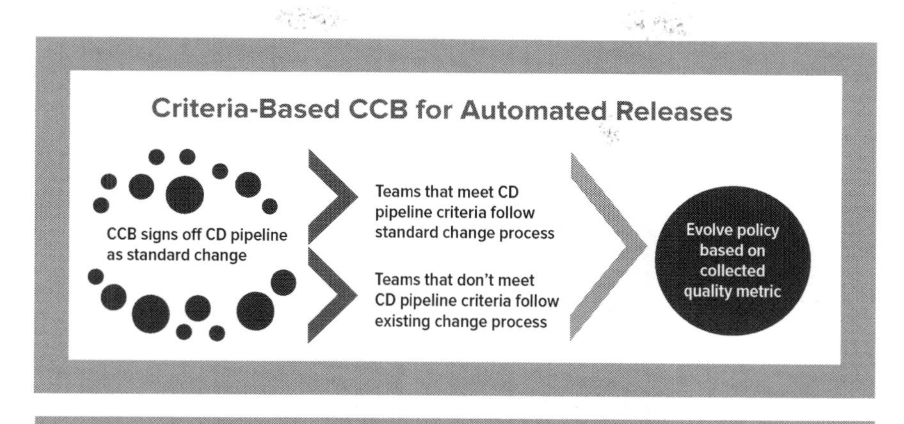

Governance Transformation

CURRENT	TO BE
Human gates in form of forums and meetings	Self service, automated quality gate
Reliance on PPT and documentation for compliance	Reliance on incremental scan of codified policy for compliance
CAB/CCB develop policy and document it for consumption	CAC/CCB develop policy and automate compliance
Forum used for compliance enforcement	Forum used for information sharing and quality audit based on compliance data
Policy evolves based on team's feedback	Continuous improvement of policy based on automatically collected data
Approval cycle measured in weeks and months	Approval cycle measured in minutes

Option B: Criteria based CCB for automated releases. The CCB signs off on the CD pipeline as a standard change. When teams meet CD pipeline criteria, there is no additional CCB review. When they don't, the change follows the existing change process. The policies are continually monitored, re-evaluated, and adjusted as needed.

Pros: Raises the bar for automation and is a model of good to other teams.

Cons: Some teams maybe left behind.

Let's get back to our original question: Do we still need the CCD with CD?

Sherry and Edward would argue that you still need a CCB, but that its role should change. Rather than reviewing each and every change, it should set the governance policy. After all, for now, you still need humans to set the policy. But maybe not by the time the 2038 All Day DevOps conference rolls around. :)

CHAPTER 8

Continuous Everyone: Engaging People Across the Deployment Pipeline

presented by Jayne Groll

CHAPTER 8
Continuous Everyone: Engaging People Across the Deployment Pipeline

We have Continuous Integration and we have Continuous Deployment, but what's continuous across all of what we do is people. Even when tasks are automated, someone wrote the automation. So, Jayne Groll evangelizes about Continuous Everyone.

Jayne is the CEO of the DevOps Institute and the author of *Agile Service Management Guide*. She describes Continuous Everyone as "about people, culture, and collaboration mapped into your value streams."

We all hear about the importance of people, culture, and collaboration, but how well are they integrated into our organizations and processes? Is it in everything we do? Is it inherently continuous?

It all starts with culture. The Business Dictionary defines culture as, "The values and behaviors that contribute to the unique social and psychological environment of an organization." As Jayne points out, culture influences how people think, feel, interact, and behave at work.

We all have a vision of a euphoric culture. For most, that doesn't reflect the reality of the current culture, which is generally Games of Thrones. Everyone wants to sit on the Iron Throne: ITIL/ITSM; DevOps; Security; Lean; Agile; Automation; and, CI/CD. Traditionally, each of these functions has operated in isolation, convinced they are the most important cog in the machine. Now, DevOps has forced us to talk for the first time in 40 years.

> The Business Dictionary defines culture as, "The values and behaviors that contribute to the unique social and psychological environment of an organization."

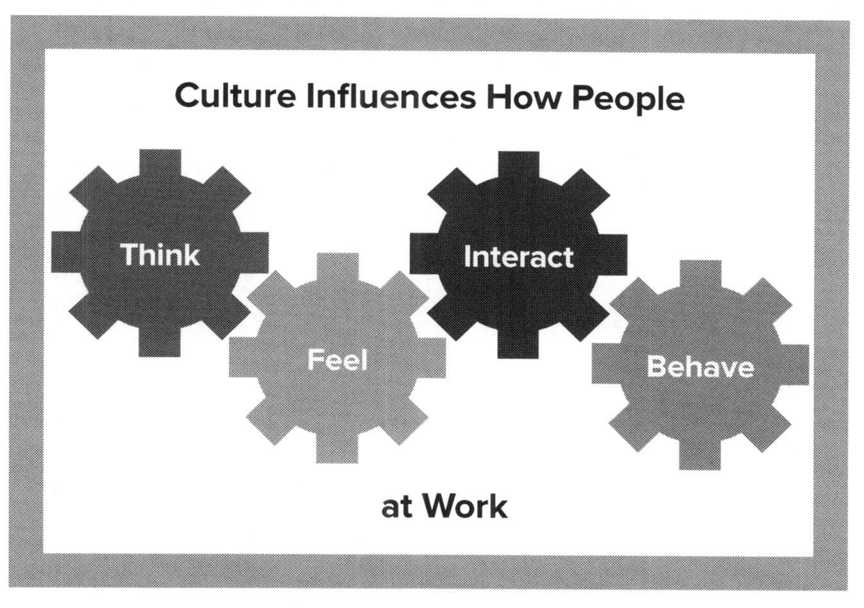

Once we acknowledge a problem, there are only 11 more steps. So, next, improving culture begins with addressing the root cause of cultural fails:

- Language barriers (industry language)
- Tribal knowledge
- Poor communication
- Thick silos
- Disparate processes
- Incompatible automation
- Inadequate collaboration

These cultural fails — both the cause and a symptom of thick organizational silos — must be broken down because they cage people's creativity, and creativity is something you can't automate.

Now, people typically participate in the Continuous People Pipeline like this: Develop → Commit → Build → Test → Deploy and APIs built in between. Jayne advocates that we need to focus on Application People Interfaces between engineers and testers and ops teams. We need to let computers do the boring, repetitive tasks and let people solve problems and focus on high value activities.

Since not everything can be automated, Continuous Engagement is critical. While some advocated for NoOps (everything is automated), we should focus on NewOps to capture the talent and expertise in operations we need to leverage for tasks only people can do, such as:

- Provide input and feedback
- Design pipelines

- Configure automation and APIs
- Review test and other output
- Design metrics
- Facilitate feedback loops
- Interface with the customer
- Solve problems
- Exchange ideas
- Be innovative

The bottom line is that this could be the beginning of a beautiful friendship. We need to reinvigorate the lost art of the dialogue. Actually talk to each other.

This begins with trust. Every aspect of the Continuous People Pipeline is filled with expertise, integrity, and people that want to learn. There is a trust. You have to trust each other's strengths and take the opportunity to engage on a continuous basis.

It is only together that we can continuously improve.

CHAPTER 9

Creating an Appsec Pipeline in a Week

presented by Jeroen Willemsen

CHAPTER 9
Creating an Appsec Pipeline in a Week

What is the key to creating a AppSec pipeline in a week? Happy developers and coffee.

Well, let's be honest — that is the key to every software development project.

Let me take you on a little adventure with Jeroen Willemsen (@commjoenie), Principal Security Architect at Xebia. He walked through his real-life experience to help the rest of us learn what to do and what to avoid.

The Challenge: Automate Security in One Week

Jeroen's client had mainframe systems for data for a physical warehouse infrastructure combined with containers on AWS and Angular, iOS, and Android native apps. Their existing workflow is diagrammed below. Joroen emphasized the importance of implementing a process like this before security automation. You need the basic infrastructure or you will fail.

The client had a wish list of new entries into the security pipeline:

- OWASP **dependency** to check third-party vulnerabilities
- License checkers to ensure licensing requirements don't expose you
- Clair vulnerability checker for Docker containers
- Dynamic analysis with OWASP Zed Attack Proxy (ZAP)
- Static analysis (they decided not to do it because it was too challenging for the week)

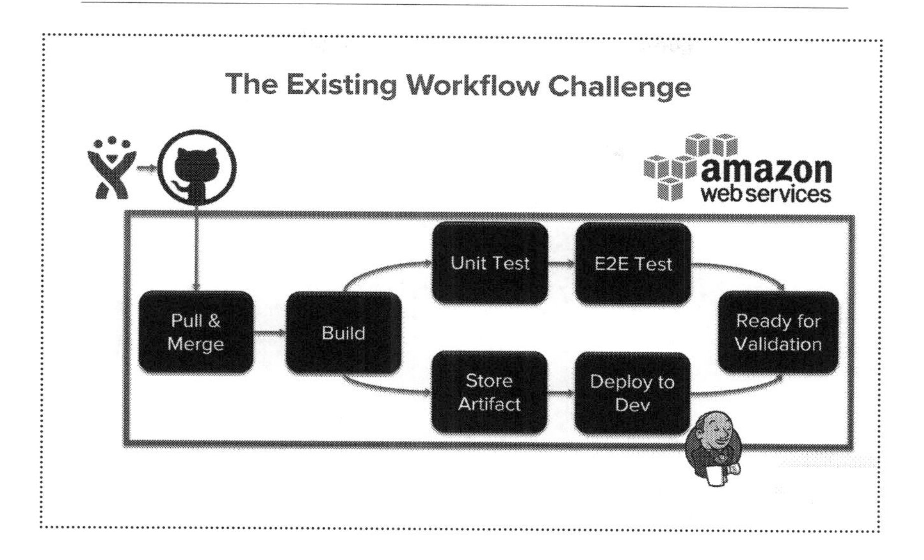

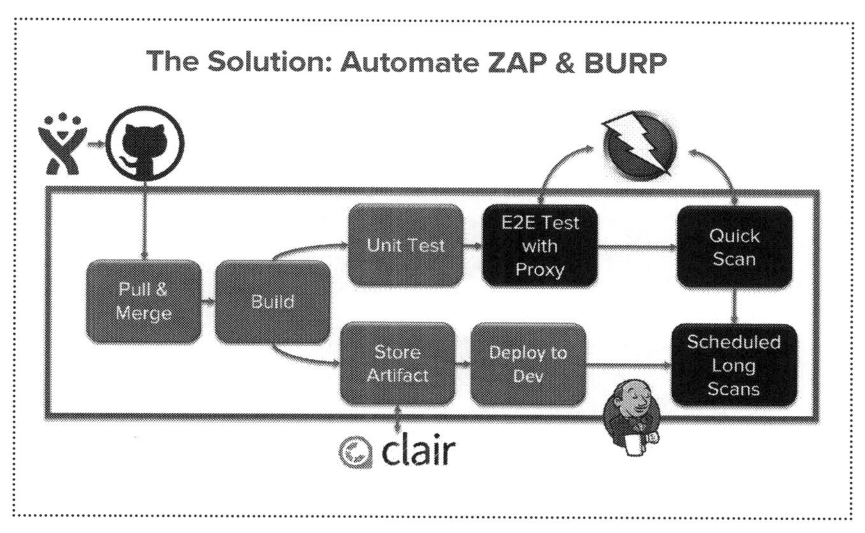

The Solution

First, they extended the build step by adding dependency and license checkers on top of quality tooling and got feedback FAST! Second, they automated ZAP. The goal? Make sure everything works in the time it takes to enjoy a cup of coffee.

Next, they had to integrate the findings from different tools. They used ThreadFix to filter out false positives and export vulnerabilities found to JIRA tickets.

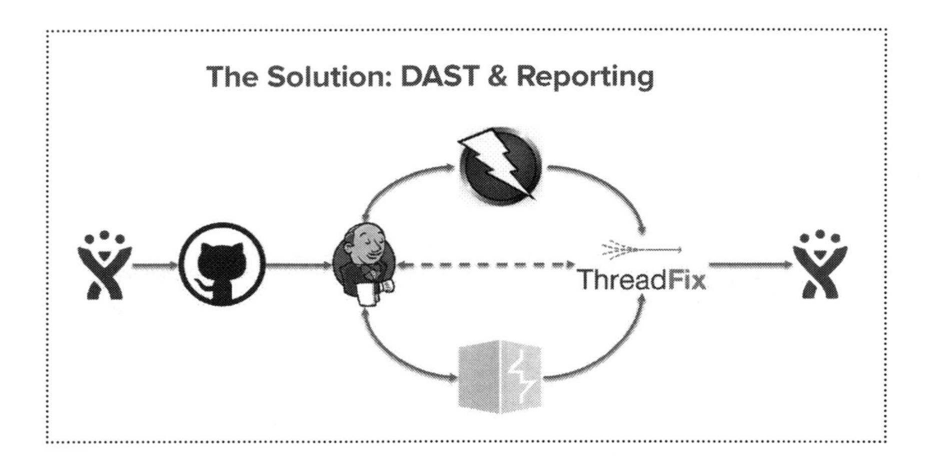

Then they ran Clair on the created containers, but they could not run Clair regularly on the registry, integrate with Threadfix, and add whitelists. Jeroen noted that this can be done now with the Clair-scanner from ArminC.

Finally, they embedded their tools in containers, gaining some key advantages:

- Less additional platform complexities
- Can run anywhere (locally/deployed)
- Easy to scale
- Still need to manage the data!
- More assets might contain vulnerabilities

However, it was not perfect — they still had to harden their assets.

Did the solution work? Yes, but it was a bumpy road.

Bumps on the Road

Bump 1: False positives. Once you start automating, you will get a lot of false positives. You have to suppress to them, but using settings and plugins or a database with a framework doesn't scale well. The third option is to have an API. This is what they did. At the time, they used ThreadFix, and now DEFECTdojo is also an option. No matter what option you choose, the lesson is the more you automate, the more you call the tools, resulting in more false positives. So, start with something.

Bump 2: Legacy APIs. Their automated tests would crash the mainframe, so they stubbed it with the help of teams. They still had to test legacy APIs separately.

Bump 3: Frustrated developers. They needed to like us to help us further. Jeroen's advice:

- Give feedback fast
- Automate all the things
- Be part of the team. Whenever you`find something, go to the developer before you put tickets on Jira, show them, and suggest how they can fix it. Be proactive
- Filter and suppress false positives ASAP. If they happen too often, developers will just find ways to work around them
- Use known tooling

Bump 4: Integrating Burpproxy. They did not complete the integration with Burp because of custom builds for containers and, at the time of testing, additional extensions were necessary to have a proper REST API.

Bump 5: False negatives. Security automation does not mean no manual pentesting. Both static and dynamic analysis tools miss context.

Bump 5: False Negatives...

Security automation does not mean no manual pentesting.

Even when you add more tools (which we have to...).

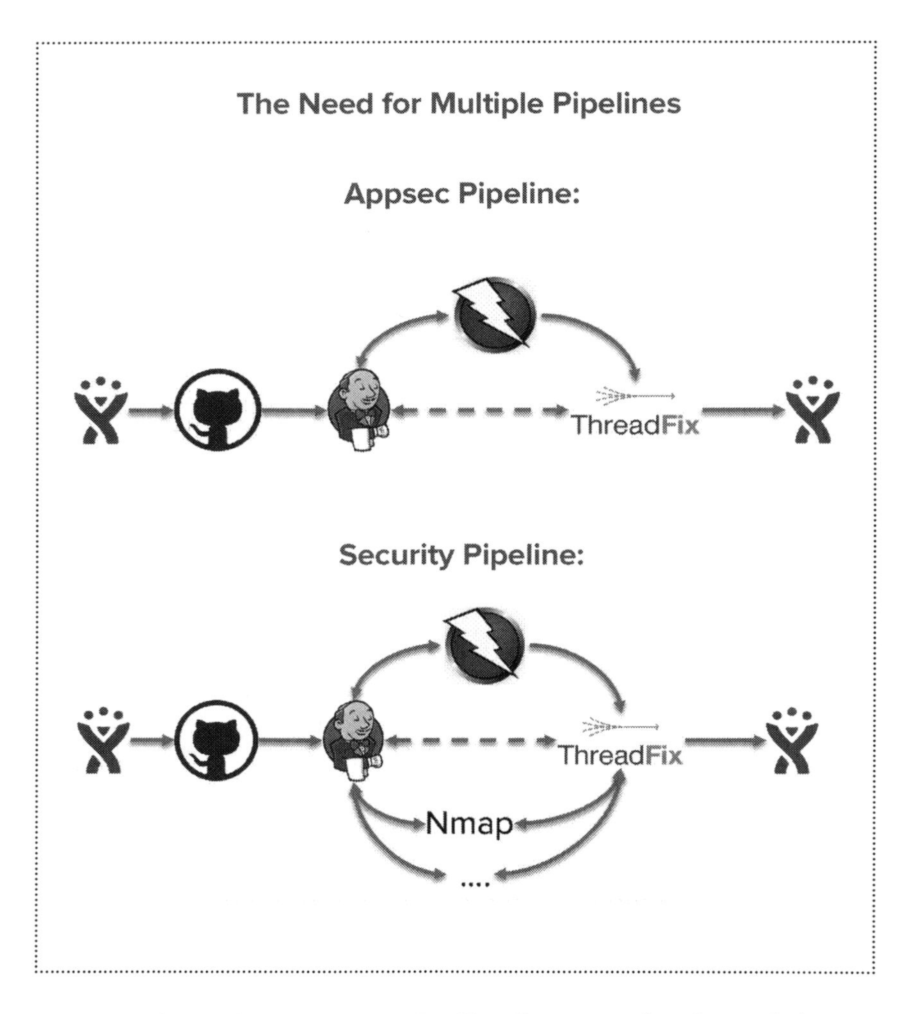

Bump 6: Platform team availability. Later on, they learned they needed multiple pipelines.

Every project and team is different, but hopefully these lessons help you be more successful when your client or product owner or manager tells you they would like an AppSec pipeline next week.

CHAPTER 10

DevOps and Security is Like Smoking Meat

presented by Apollo Clark

CHAPTER 10

DevOps and Security is Like Smoking Meat

It isn't everyone who thinks, "Doesn't Ubuntu remind you of wild boar?" Or labors over his pit of slow roasted pork shoulder while contemplating containers or dwells on e2e testing while mesquite smoke permeates spareribs.

But, Apollo Clark (@apolloclark) does. Apollo is a foodie of smoked meats (is that a meatie or smokie?) — working to master the craft — understanding different cuts, what each type of wood adds, and the subtleties of sauces. But, alas, it is a hobby — security and DevOps are his career.

Being passionate about both, he naturally sees parallels between the two.

Debian = pork, simple and delicious

Ubuntu = wild boar, a little more complex

Redhat = lamb, costs a bit more

Fedora = beef, everyday

CenOS = chicken, simple and nice

SUSE = duck, exotic and very tasty

Gentoo = squab/pigeon, a little different

Arch = Cornish game hen, kind of like chicken

Understandably, you are asking yourself, "What does smoking meat have to do with DevOps and security?" Apollo notes both have tremendous complexity and nuances, there are multiple ways of getting the job done, lots of ways to mess it up, a couple of ways of doing it right, and you are always learning.

For both, there are many tools and processes to get the job done. For smoking, your wood is a critical component, along with time. Let's look at some parallels:

Oak = unit testing

Maple = coverage

Apple = dynamic analysis

Peach = static analysis

Mesquite = e2e testing. People love it, but it is difficult to handle

Wine barrel = browser support. This is when you are doing it really well

Bourbon barrel = device support. Pretty complicated.

Smoking time = testing. You can test or smoke for 5 min or 5 hours, but there is a sweet spot. You can over-smoke meats and you can over test.

Getting Hungry?

Meat is critical to a meal of smoked meat, but so are many other components. Likewise, software applications take a suite of other tools and components to deliver the full package. Get your taste buds ready, because Apollo likens all of the goodness that goes with smoked meats to tools you use to keep applications running:

Sauces = auto-scaling. They both can get pretty complicated. Always a little different, but you have to make it work for you.

Bread = monitoring, which is the bread and butter of infrastructure. Make sure you have it up and down your stack.

Salad = system logs. Not the sexiest things, but you can rely on them.

Fruit = custom application logs. Takes a lot of time to pair them, but takes a really good thing and makes them better.

Beer = firewall. You should always have both.

Wine = Intrusion detection systems. Gets better with time.

Whiskey = IR training. It takes time and there are so many ways of doing it. When things break, we have procedures on how to deal with them.

When smoking meats and in DevOps and security, Apollo asks and answers, "Do we have to do everything? No, but the more we do, the better it will taste."

"Does it cost money and take time? Yes, but you can do great things even without money and time. You can use cheap meats and you can use open source. Start simple, build up complexity, and always be learning."

Constantly ask yourself, "Am I better today than I was yesterday." If not, be better.

CHAPTER 11

DevOps for Normals

presented by Michael Coté

CHAPTER 11
DevOps for Normals

Do you want to know why we do DevOps but are neither Dev nor Ops?

Michael Coté's (@cote) talk is for you. He is the Director, Marketing at Pivotal, and talks about DevOps for people who want to know what the big deal is.

First, Michael answers the question, "How did we get here?" Of course, it is a complicated answer, but it really boils down to the need to innovate faster and better. He quotes Stephen Bird, CEO of Citi Global Consumer Group, "In order to grow Citi, we first have to grow our own perspective, skills, and capabilities… Our curiosity, our openness to learning and trying new things, our ability to adjust and adapt quickly, and our willingness to fail fast and fail small are the essence of a culture that innovates and exposes new value to our clients in real time."

DevOps is about improving software development — being more lean, agile, efficient, and able to adapt to changes and improve quickly.

"Since 2000, 52% of the names on the Fortune 500 list are gone, either as a result of mergers, acquisitions, or bankruptcies."

— R. RAY WANG

Expected BigCo Lifespan

1960s: 60 years on S&P 500

2020s: 12 years on S&P 500

It rewards a culture of innovation and effective communications. It asks, "What is the business reason for this technology decision?"

While many companies are transforming their digital side, many are lagging behind. Quoting one research study that looked at companies' digital transformation strategies, only 40% were at any stage of a formal strategy. Another looked at an alarming trend of IT's role in business innovation, showing a downward trend — from 56% in 2013 to 31% in 2015 — of involving IT.

While many companies have a long ways to go, many others see the need to innovate in technology. One wave is the move to Agile software development and DevOps.

At the core of DevOps (and Agile) is to use a "small batch approach" at all levels. As Michael states, this means, "You are continually improving because you have no idea what you need." The small batch approach scopes projects down to a week, maybe two weeks, or a month, as opposed to 9 months or more.

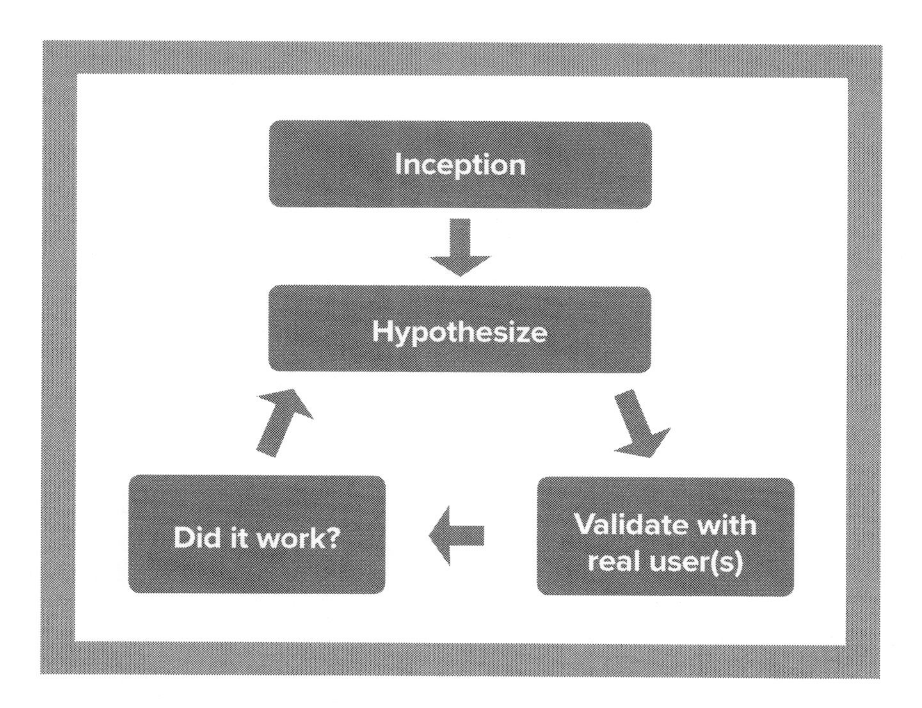

Michael contends, while the scope and timeline are much shorter, the reality is it is really the waterfall method, just in a shorter time frame. It has the same activities, but just increases the iterations. This gives you more data to ensure user-centric design and improve the software, and allows you to adjust quickly based on real-time feedback.

We saw earlier that there is a still a long ways to go in the digital transformation of organizations.

Part of the challenge is scaling beyond the team level. The bigger the organization you work in, the more you need to pay attention. And, you need to stop hitting yourself. Instead of saying, "We can't do this," decide you need to change and get to work. Michael calls this the anti-pattern. This starts with management, and, like Dev and Ops, Management need to take a small batch approach and accept and take responsibility for failures. This gives management a lot more visibility and transparency.

The bottom line is that great organizations don't just happen, and they don't have eternal life. They must innovate, and, in this day and age, have to do it quicker and better than their competition. The principles of DevOps are part of that.

CHAPTER 12

DevOps for Small Organizations: Lessons from Ed

presented by Ed Ruiz

CHAPTER 12
DevOps for Small Organizations: Lessons from Ed

Ed was demoralized. He had just heard a speaker who would change his life. He knew he needed to change, and he knew what the end goal was. He just didn't know how to get there. He needed fresh air. He needed endorphins. What better way to do that than go on a 6-hour run through some of the seedier neighborhoods of Vegas to the edge of the desert.

Ed Ruiz (@eruiz06) is the CIO for the Association of Schools and Programs of Public Health (ASPPH), and his DevOps journey started in the desert outside of Las Vegas. Earlier that year, ASPPH recognized a need to modernize their membership structure. They had ballooned from 31 to 106 members. Ed had a staff of 10, supporting 54,000 students and 13,000 faculty serving in 141 countries. They were steeped in following the Microsoft Enterprise handbook, deploying via the waterfall method every 3-6 months, and recognized they needed to update their IT infrastructure.

There, amongst the glitter and glam of a ballroom on the Strip, Ed heard DevOps Evangelist Andi Mann (@AndiMann). Ed was glued to what Andi had to say. This — this is what our organization needs. But he quickly became demoralized because he didn't know how to lead the drastic change.

During Ed's run following Andi's keynote, he realized he needed to start with a vision. Spoiler alert — Ed was successful, but not without a few hiccups. Ed shared four lessons he learned during the session to help the rest of us pursuing our DevOps journey.

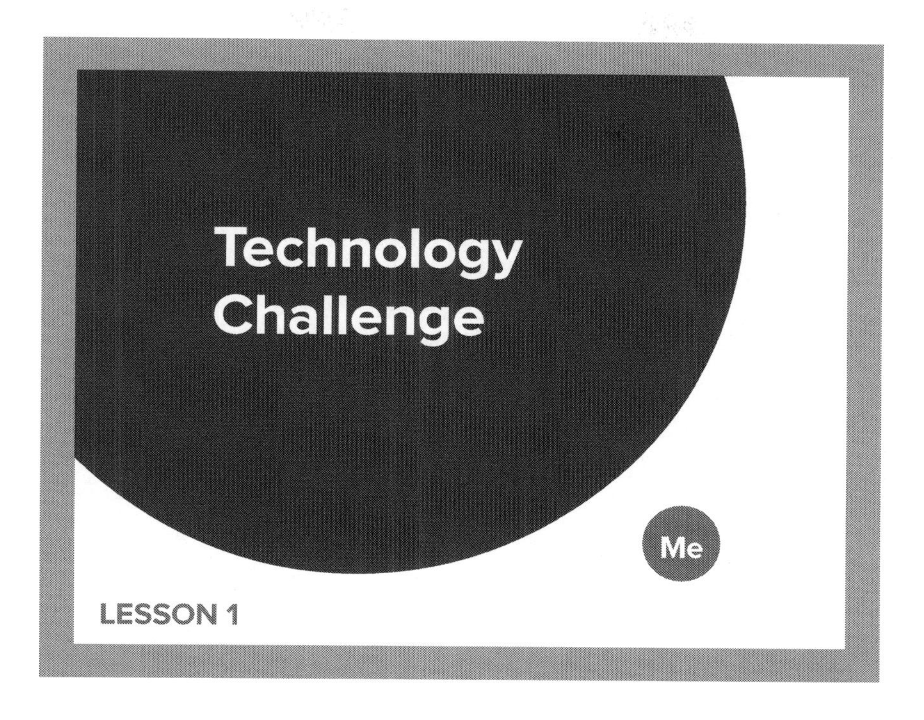

1. Articulate a Vision.

Your team needs to understand what the organization does, why it is important, and how this change allows the organization to be more effective.

2. Start with Culture and Sharing.

The DevOps mantra emphasizes the equal importance of Culture, Automation, Measurement, and Sharing (C.A.M.S.). Too often people are drawn to automation and measurement first. It is what they know. It is the DevOps tools that are fun to implement. According to Ed, over-emphasis on tools is a huge mistake. DevOps is called DevOps because it is implementing a culture of sharing between development and operations. Automation and measurement are tools to facilitate sharing, but if your team isn't on board with culture and sharing, DevOps will fail.

3. Prune the Bad Apples.

We all know that person (they go by many names: Dr. No, Negative Nelly, Milton). You don't want that detractor in your organization. When Ed decided to keep a Dr. No on his team because of his technical acumen, it took half of Ed's team quitting before he realized the bad apple needed to be pruned from the tree. Dr. No's negativity was contagious and led to poor morale that was eroding the culture needed for success. Ed's advice? You can teach skills, but you can't teach a positive attitude. Hire for the attitude first.

4. Leadership Buy-In is Essential.

Priorities and budgets will change and setbacks will happen. When you have buy-in from the leadership, it creates room to fail. You can pick yourself back up and take another step toward your vision.

We don't know what else happened in Vegas over those five days, but we do know Ed came home a changed man, he changed his organization, and we can all take some lessons Ed learned and apply them to our own transition to DevOps.

CHAPTER 13

DevOps: Building Better Pipelines

presented by Dan Barker

CHAPTER 13
DevOps: Building Better Pipelines

"The deployment pipeline takes value from development and delivers it to clients."

This is one of the better summations of a development pipeline that I have heard, and it came from Dan Barker (@barkerd427). Dan's talk focuses on the value of pipelines, how to build them well, and how to build both development and infrastructure pipelines.

Traditional Pipelines

The traditional pipeline: development-->test-->production servers is a good foundation, and, like the plumbing in your home, is helpful to generally get us where we want to go. However, as anyone who has raised boys knows, a toilet doesn't guarantee 100% compliance. The user isn't perfect.

It is similar to basic development pipelines. Things don't go quite as planned. As Dan mentioned in his talk, as a developer, he strides to keep his code all in dev, and then push to test, and then push to production. However, that doesn't always happen. And there are other inconsistencies that inadvertently crop up (such as tab vs. space), despite everyone's best intentions.

Dan is a chief architect and works in highly-regulated industries — financial services, health care, and insurance — so precision, compliance, audit trails, etc. are all the more important. He architects pipelines to ensure human error is minimized.

What's in a Pipeline?

But why all of the fuss over pipelines? Well, first Dan touched on some of the values pipelines provide:

- Abstract audit and compliance
- Trivialities eliminated (spaces vs. tabs)
- Security checks occur early/often
- Tests all the things
- Nimble security
- Common artifact repositories
- Standardized approval system
- Apps become secure by default

But how do you architect a pipeline to deliver these values?

Dan offers key points that all of us should consider applying:

- Production data is separated from development data. You can only get there through the pipeline
- You interact with the Platform, and it configures the levels below it
- You can duplicate containers as much as you want to scale horizontally, and can replicate that in production
- The configuration must come from the environment

As he dug in, he started with covering the advantages of two types of pipelines, recognizing you need to select this first:

Scripted:

- Very Groovy!
- More powerful
- Provides greatest level of flexibility

Declarative:

- Only a little Groovy
- Simpler to maintain
- Easier to read and understand

Dan then dug into the technical nitty-gritty, such as shared libraries, setups, software to manage pipelines, fabric8, Kubernetes, Yahoo! Screwdriver, etc. and then made the observation that he thinks that deployment pipelines have fallen behind. In his real-life example, he and his team decided to take a more holistic approach to their pipeline (think: the first way of DevOps) so they could build and deploy apps and also build and deploy the infrastructure.

Sharing Knowledge

They built a system that is especially applicable to those in highly-regulated industries, and they hope to open-source it soon, so keep an eye on @barkerd427 for more info.

CHAPTER 14

DevOps: Escape the Blame Game

presented by Matthew Boeckman

CHAPTER 14
DevOps: Escape the Blame Game

We have all been there in a postmortem when someone says, "Let's get to the root of the problem." And, we all know what that means: who or what is to blame?

We also all know that no one wants to play the blame game, yet we all do. But it isn't our fault (no blame, see what I did there?). It has been the default system for solving problems in business for decades. It is called root cause analysis (RCA).

We can change — for the better.

Matthew Boeckman (@matthewboeckman) is SVP of Engineering at Bluprint. He grew up a systems guy and jokes that he has been in DevOps for 18 years, even though DevOps wasn't around, because he has always been nice to developers.

Digging in (pun intended), RCA focuses on what went wrong, and how we can prevent it from happening again.

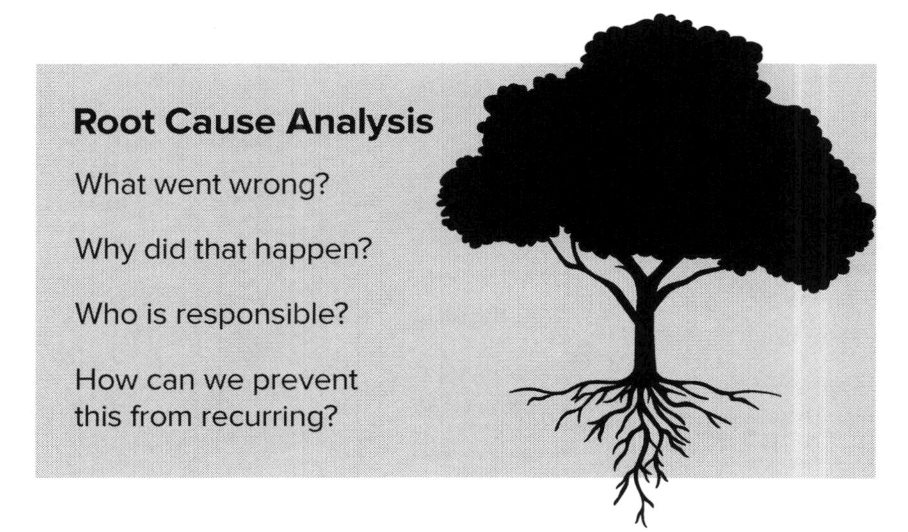

The core problems with RCA for development is that it doesn't provide for enough complexity and its natural focus is blame, which can undermine a positive DevOps culture.

RCA was more applicable when waterfall was the development methodology because states stayed consistent for months or even years at a time. In the age of Agile, DevOps, CI/CD, microservices, etc., states of work are in a constant flux. RCA can't provide solutions quickly enough. As Matthew notes, in RCA, things are either good or bad, working or broken, uptime or failure. The reality is that our world is more nuanced.

What Matthew recommends is to look at it through the principle of emergence because it, "separates judgement from the good and the bad binary approach to our system health, and instead focuses on behaviors and interactions, patterns and complexities of our system. With practice and effort we can manage them to more desirable states." But what does this look like in practice?

Getting back to the analogy of the tree and its roots, the answer is more of a forest than a tree. Trees are one living organism, forests are ecosystems.

Matthew takes this philosophy and mental picture and gives us a better system — Cynefin. It is a Welsh word that means habitat, and was created by Dave Snowden (@snowded), originally for managing IBM's intellectual capital. It draws on research in systems, complexity, network, and learning theories.

Starting in the bottom right quadrant, working counter-clockwise, its goes from simple to more complex.

Simple

These are patterns or behaviors that don't require a great deal of understanding. DevOps is increasingly setting up automated systems to respond to simple issues.

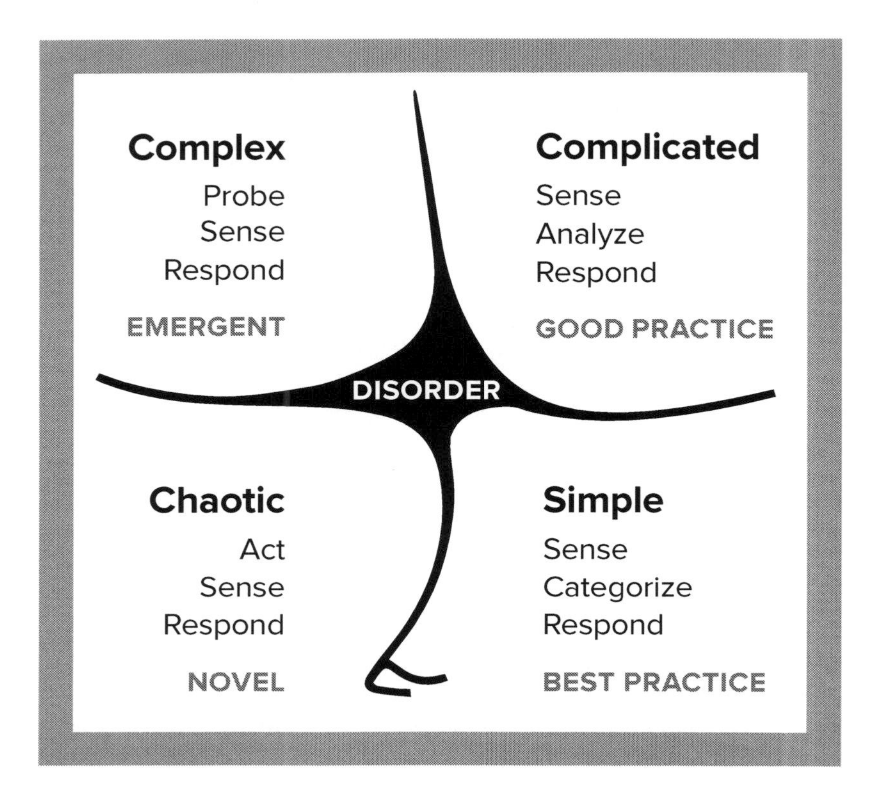

Complicated

These are known unknowns. You can imagine a set of realities where they can occur, and they are probable, but not certain. For instance, a busy harbor might get a storm that causes damage to boats, docks, etc. It is hard for the harbor manager to manage and they need to think about it. This requires people to do some thinking, and it is difficult, if not impossible, to automate.

Complex

This is where we start to see emergent behaviors occur. We don't have the metrics need to understand or manage these problems or you haven't looked at that metric before. We start with probing, going into the system, and exploring. Think of any collection of humans at any scale. Things are still in the scope of probable, but things change quickly. There are many moving parts that aren't predictable and that we didn't fully encounter in our test methodology.

Chaotic

This is, well, chaos. Matthews' real-world example was an entire region for AWS went down, causing other regions to be overloaded as system admins were moving services. In chaos, you *act*, then get a *sense* of where things are, and then *respond*.

Disorder

In DevOps, this is where you have lack of communication and collaboration. Here teams need to: *reduce*: figure out what you agree on; *analyze*: build consensus; and, *iterate*: move to a quadrant and continue.

Matthew notes that knowledge and practice move patterns towards more favorable quadrants. But, complacency erodes the process. Complex systems left poorly managed will create increasingly complex processes to manage.

How to Adopt Cynefin

- In the moment: Ask, what quadrant does this map to?
- In the post incident report: How did we manage the pattern? Was it complicated, complex, simple? What can we do to change it?
- In your sprint planning: Devote time to manage your patterns clockwise. What can we move with a little bit of work?

The reality is that RCA is really only present after the fact. Cynefin is always present, dynamic, expects change, and calls us to action — sounds a lot like DevOps.

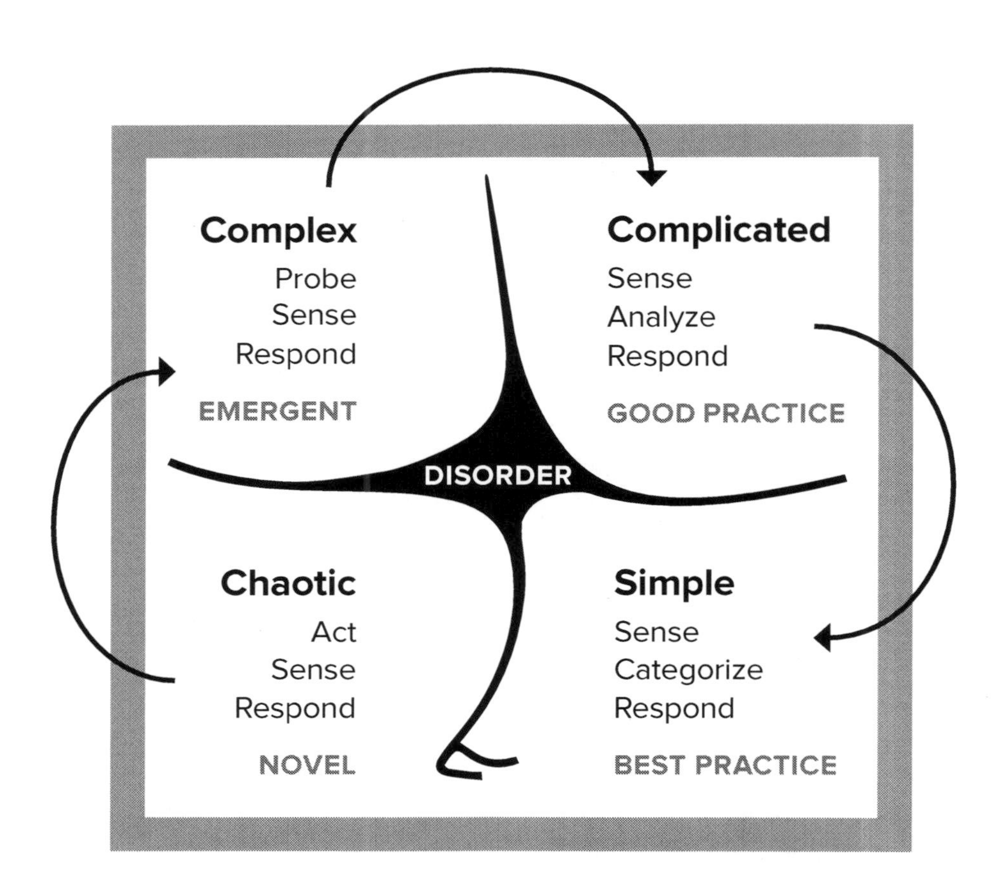

Root Cause Analysis	Cynefin
Simple Causality	Dynamic
Static Model	Expects Change
Binary Thinking	Embraces Emergence
After-Action	Present in the Moment
Focus on Blame	Call to Action

CHAPTER 15

DevOps: Making Life on Earth Fantastic

presented by Helen Beal

CHAPTER 15
DevOps: Making Life on Earth Fantastic

DevOps is taking over business. Not because technology permeates business, but because it has broadened to include the entire business value stream. Different best practices throughout the enterprise are incorporating principles of DevOps to deliver better outcomes to customers.

Helen Beal, DevOpsologist at Ranger4, spends her days "making life on earth fantastic" in part by helping implement DevOps philosophies in organizations.

Helen pointed out that DevOps is about 10 years old, yet, unlike Agile, there isn't a DevOps Manifesto. But, we know what DevOps is about. She quotes Mark Schwartz from The Art of Business Value, "DevOps, in a sense, is about setting up a value delivery factory — a streamlined, waste-free pipeline through which value can be delivered to the business with a predictably fast cycle time."

Helen also introduced the audience to the DevOps equivalent of the OODA loop — Ideation → Integration → Validation → Operation → Realization → and repeat, and the CAMS principle: Culture; Automation; Measure; and, Sharing. A process and principles that can be applied inside and outside of IT and software development.

She is leading up to making the point that business best practices systems are converging around DevOps — a concept she calls the Emerging DevOps Superpattern.

These are best practice systems — some of which have been around of more than a half-century and others that aren't even a decade old — where, as they mature and evolve, it is becoming evident they share best practices with DevOps. DevOps is at the center of improving business.

The DevOps Loop

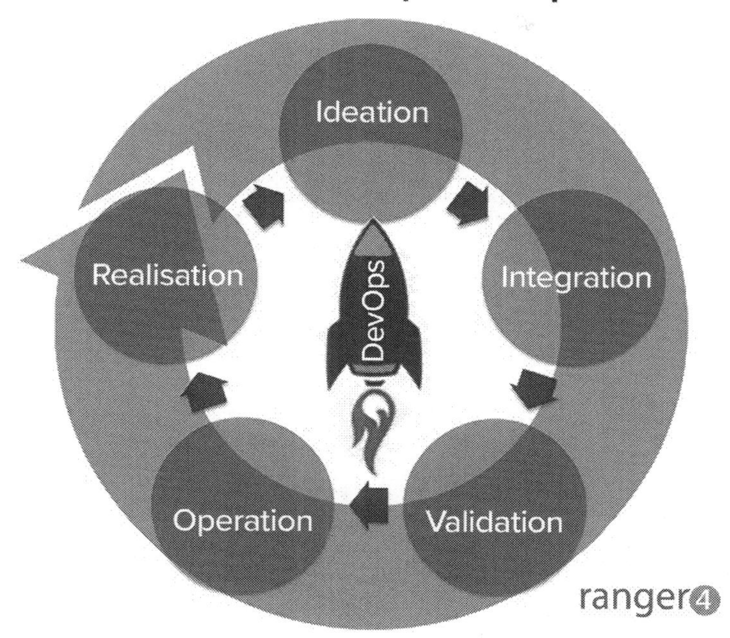

The Emerging DevOps Superpattern

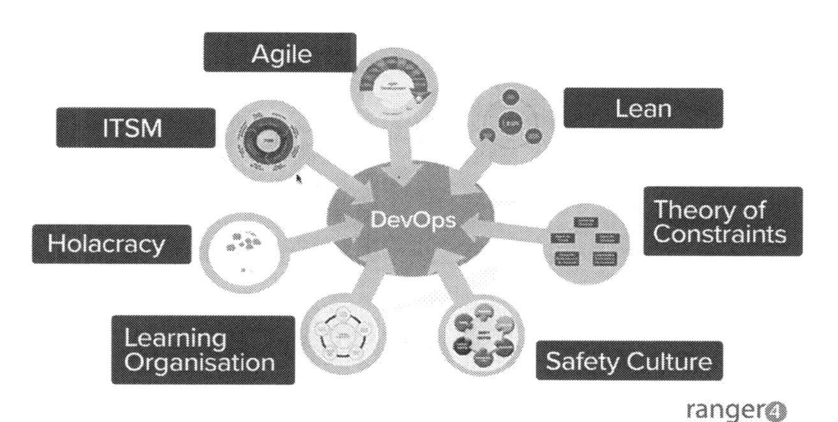

Helen looked at each best practice system and principles they share with DevOps:

Agile: Support and trust are key; the first principle of the Agile Manifesto is continuous delivery; measuring value to the customer; daily collaboration across functions.

Holacracy: Everyone has the ability to call out when they see a problem; heavily focused on using peer-review processes and relies on collective intelligence.

ASM (Agile Service Management): This builds on ITSM. It is just enough governance to deliver the best service to the customer; promotes better collaborations by cross-pollinating vocabulary and methods.

Lean: Focuses on delivering value to the customer with minimal waste; the types of waste Lean seeks to eliminate are errors and duplication — both of which automation helps to tackle; uses Value Stream Mapping to understand the handoffs between processes and human interactions.

Learning Organization: Decentralizes the role of leadership; puts long-term sustainability ahead of short-term fixes; automates rote tasks to release time for learning and experimentation; uses knowledge management tools; touts exposing personal mental patterns and thinking for inspection and influence from others.

Safety Culture: In a highly experimental, innovative environment, we need to build safety in. Fail safe, fast, smart — testing and auditing early in the release cycle and pre-emptive monitoring; Mean Time to Repair but measuring failure in terms of real business value; accountability and ensuring all understand their role in procedures is key. In DevOps we love failure because it shows we are innovating.

Theory of Constraints: Mental models held by people can cause behavior that becomes a constraint; automation can remove constraints in manual processes; constraints are frequently poor handoffs due to weak collaboration.

The reality is DevOps embraces principles that make business better — better for the business, for the employees, and for the customers. That is seen as other systems, outside of software development, embrace its principles to improve business.

CHAPTER 16

DevOps: Making the Boring Things Stay Boring

presented by Mykel Alvis

CHAPTER 16

DevOps: Making the Boring Things Stay Boring

Mykel Alvis' (@mykelalvis) likes to say, "I, For One, Welcome Our New Robot Overlords." He isn't trying to curry favor with the new robot rulers, ala Kent Brockman, but, instead, evangelizes on the importance of precision in DevOps.

Mykel is the DevOps Coach at Cotiviti Labs, the development arm of Cotiviti. Every day they handle healthcare data, which as you know happens to be some of the most regulated and protected data around. As Mykel says, "We don't do cat pictures on the Internet." A mistake, breach, security flaw, etc. could mean the exposure of privacy and financial data for millions of people. But, they are a large organization, with exposure that needs to be managed. They have 200-plus repositories, 3-4 deployments per day, 15 developers, and 0 operators (sort of). Without the right discipline and rigor, costly mistakes could happen.

Hence, Mykel's need for the robot overlords. As he notes, robots have, "Lower utility, but greater predictability." They do exactly what they are told, and they do it the same way, every time. By contrast, humans are superior to responding to unpredictable events. Mykel's goal is to leverage the strengths of each. Cotiviti Labs has systems, policies, procedures, and a culture in place that supports the automation of tasks as much as possible to ensure mistakes are kept to a minimum and the humans can focus on the unpredictable. They are striving to make the boring things stay boring. They call it "The Labs Way" and here is how they do it:

- **Treat everything as code.** Everything. Infrastructure, data, everything. As an example, they perform pull requests to get status updates for executive reports.
- **Test the code.**
- **Perform formal releases.** They build their code in very specific ways, with extremely rigid gatekeeping processes.
- **Releases produce immutable artifacts.** Once something has been released, they will never change it. They could, but they don't.
- **Keep everything.** Not in a huge, unorganized file system. In an organized repository.
- **Failing tests mean a failed build.**
- **Design your systems to be automated.** They don't do things that can't be done automatically.
- **Deal with those systems only through automation.** This is probably the most challenging thing we do since it is extremely hard to convince developers to write their code to do this.
- **Operations IS developing.** Operations is not a bucket for development. Operations is a development task. You treat it like it is real code.
- **Because everything is code.** Everything.
- **All defects are defects of code.**

Treat everything as code. Everything.

To achieve this, they:

Standardize

1. Evaluate
2. Test
3. Discard the useless
4. Own it

None of this comes easy. It challenges norms. It pushes boundaries. It mandates discipline. Your team has to understand you have a way to do things, there is a reason, and it will reduce risks and make sure everything works. It will keep the boring, boring, and that is exactly the goal.

Standardization is an effect of applying discipline to manufacturing.

Suffer the pain of discipline to manufacturing or suffer the pain of regret.

CHAPTER 17

DevSecOps: Overcoming the Culture of Nos with Chaos

presented by DJ Schleen

CHAPTER 17
DevSecOps: Overcoming the Culture of Nos with Chaos

Traditional security has thrived in a culture of "no."

The Culture of "No"

We have all met that wall. And when those walls exist, people find ways around them. The workarounds make their lives easier. They implement what they think is best. Their efforts are not intentionally destructive, but can lead to unintentional vulnerabilities and, potentially, harm.

While DJ Schleen (@dschleen) studied physical architecture and design, he now works in software security architecture and is a DevSecOps Advocate at Sonatype and a former Security Architect and DevSecOps Evangelist with Aetna.

The culture of "no" is exactly the kind of culture DevOps is designed to improve, and, as DJ asserts in his talk, "DevOps is an unprece-

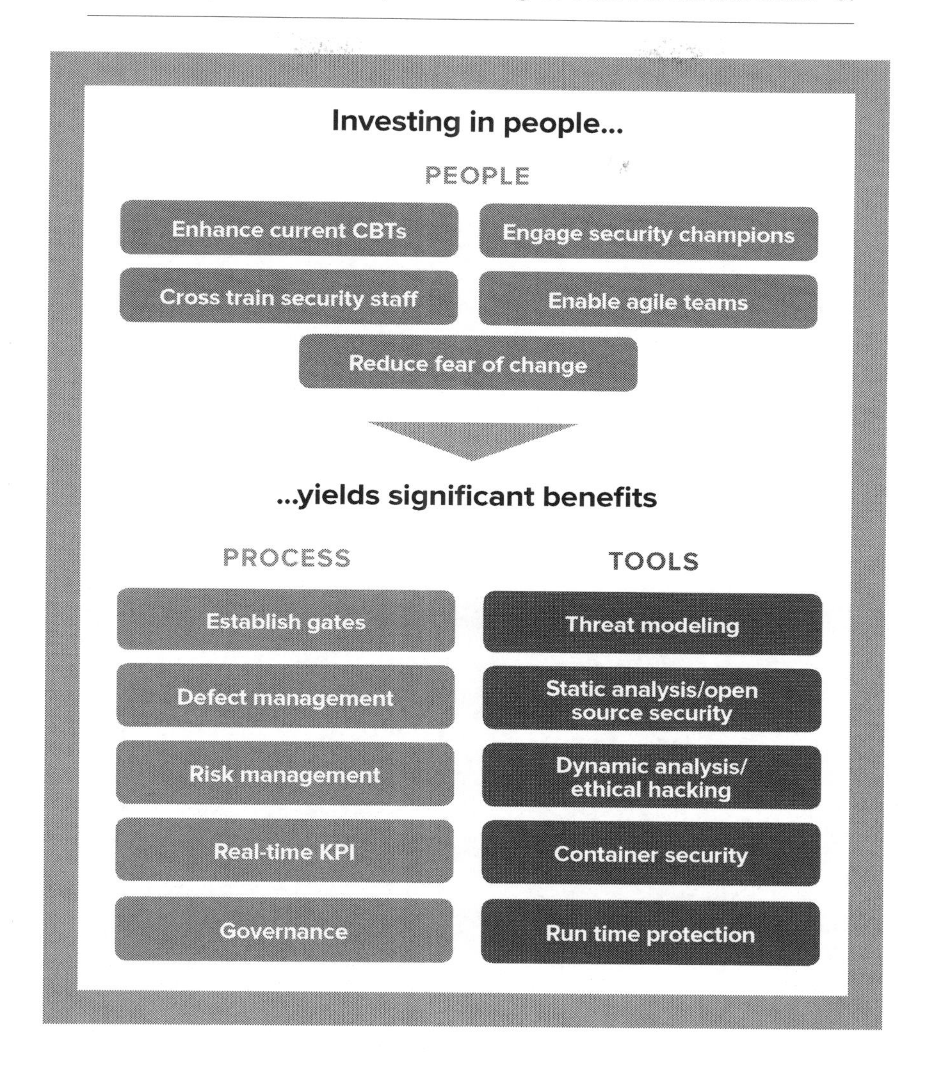

dented opportunity for security." DevOps provides a system to react quickly by supporting a continuous delivery culture and the addition of security controls into an automated environment.

Invest in People

DevOps is also about investing in people, improving the lines of communication between development, operations, and security, and automating where you can automate to give humans the ability to focus on what we do best. You maximize success with DevOps when you invest in people, which, in turns, also improves your processes and tools.

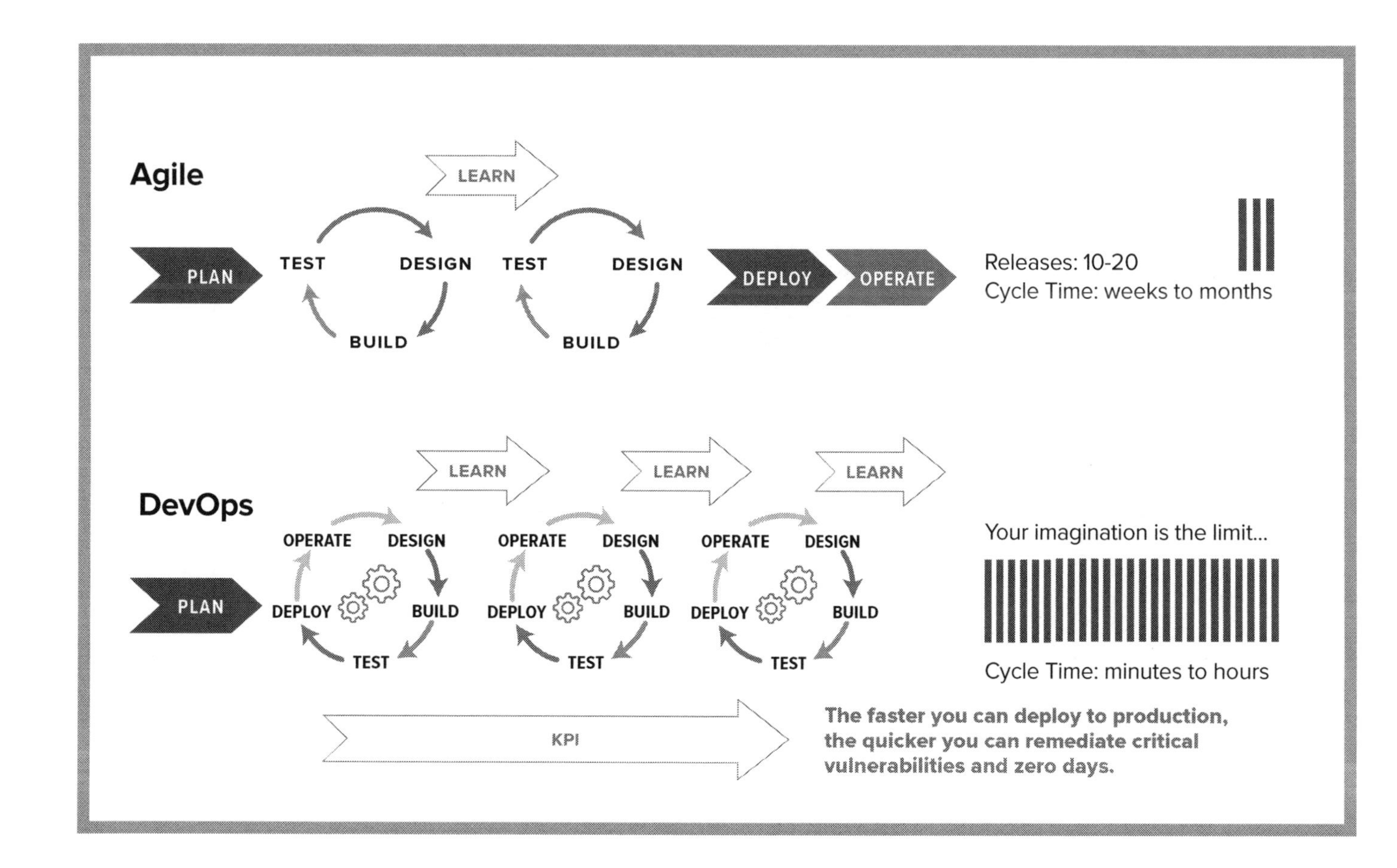
Agile
LEARN
PLAN
TEST
DESIGN
BUILD
TEST
DESIGN
BUILD
DEPLOY
OPERATE
Releases: 10-20
Cycle Time: weeks to months
DevOps
LEARN
LEARN
LEARN
PLAN
OPERATE
DESIGN
DEPLOY
BUILD
TEST
OPERATE
DESIGN
DEPLOY
BUILD
TEST
OPERATE
DESIGN
DEPLOY
BUILD
TEST
Your imagination is the limit...
Cycle Time: minutes to hours
KPI
The faster you can deploy to production,
the quicker you can remediate critical
vulnerabilities and zero days.

So, how do we change security from a culture of "no" to a culture of "yes" — or, perhaps more appropriately, "Yes, but this is what it looks like."

To start, DJ first looks at the underlying system and asks, "Is Agile 'agile' enough" to force this change. When his answer is answer "no" he knows it's time for DevOps.

Too often, Agile is just a collection of mini waterfalls. DJ states, "DevOps breaks the chain of waterfalls. With DevOps you can get fixes out quickly and easily. No one has to come in on Saturday."

Building Security into the DevOps Pipeline

To fully recognize the benefits of DevOps for security, DJ notes you must build security into the pipeline, automate wherever possible, and have security professionals code too. DJ outlines some goals:

- Be a culture of yes and working together
- Deliver remediation guidance back to developers
- Integrate security knowledge and secure coding practices into your DevOps teams
- Have security teams "get hands dirty" by coding
- Don't just automate the scan button
- Pass software through well-defined and automated gateways where guardrails are in place to assess code security without decreasing velocity
- Utilize automation to stop or pause the delivery pipeline when critical vulnerabilities are detected or manual intervention is necessary
- Utilize automation to provide actionable remediation guidance
- Remember software after production deployments

Chaos in Your Comfort Zone

What are some practical techniques DJ recommends? First, introduce chaos.

Chaos is a matter of stretching your comfort zone. DJ recommends going beyond traditional monitoring. Consider:

- Randomly take containers down
- Explore holes in your development practices
- Unleash random hell on your infrastructure and applications
- Continuously review the stability and resiliency of your systems
- Execute attacks in production

DJ digs into the details of each of these techniques, but, in the end, it all comes back to introducing chaos to improve your preparedness for attacks and your understanding of the entire system.

Of course, this approach has challenges. DJ mentions selecting the right tool sets, tailoring the people, processes, and tools to unique environments, teaching old dogs new tricks, having multiple "flavors" of DevOps, and making KPIs and indicators actionable.

Don't Go the Road Alone

DJ's key takeaways are:

- Don't fear deploying rapidly and often into production
- Always gather information in the form of KPIs and make them ACTIONABLE
- Support the organization with tools, techniques, and best practices
- Automate EVERYTHING
- Defects are defects — regardless if they are a code defect or a security vulnerability
- Code your infrastructure — eliminate access to physical or cloud-based machines
- Choose tools that interfere minimally with flow
- Introduce chaos to become a moving target

But, broadly, he also emphasizes not going the road alone. We all learn from one another, so listen, ask questions, and give back where you can. Oh, and get to a culture of yes.

CHAPTER 18

Docker Image Security for DevSecOps

presented by José Manuel Ortega

CHAPTER 18
Docker Image Security for DevSecOps

Docker. It seems like in this day-and-age you are either using Docker containers or you are going to use Docker containers. If you are on the bandwagon or are thinking about it but have concerns about their security, it's time to read on.

José Manuel Ortega (jmortega.github.io) is a software engineer and security researcher in Spain. His talk gave an overview of typical Docker deployments, explained the attack surface and threats, presented how to detect vulnerabilities, and outlined a couple of best practices. In short, his advice will help you learn how to better secure your Docker containers.

[New to Docker? Read this paragraph; all others skip ahead.] José offers this explanation of what Docker is, "Docker containers wrap a piece of software in a complete file system that contains everything it needs to run: code, runtime, system tools, system libraries — anything you can install on a server, regardless of the environment it is running in."

That is, containers are isolated but share an operating system and, where appropriate, binaries and libraries. Docker provides an additional layer of isolation, making your infrastructure safer by default. This makes the application lifecycle faster and easier to configure, reducing risks in your applications.

For starters, José lays out Docker's default mechanisms for security:

- Linux kernel namespaces
- Linux Control Groups (cgroups)
- The Docker daemon
- Linux capabilities (libcap)
- Linux security mechanisms like AppArmor or SELinux

José walks through others tools, add-ons, best practices, etc. to increase Docker container security. I will cover most of them here.

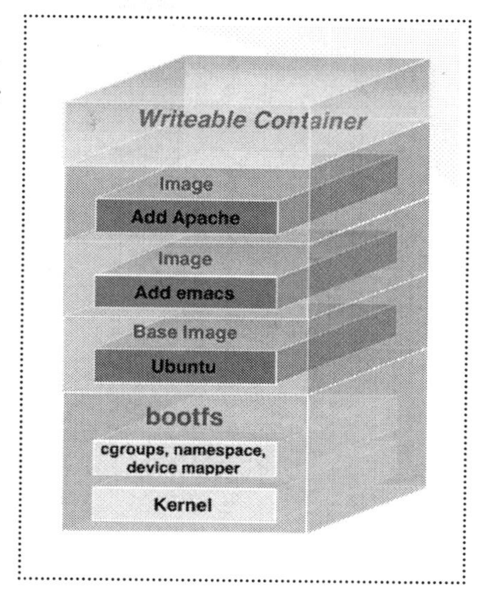

Docker Inspect Tool. The Docker Inspect Tool is built into Docker. It provides information about the host name, the ID of the image, etc. and it comes up when you start Docker.

Docker Content Trust. It protects against untrusted images. It can enable signing checks on every managed host, guarantee integrity of your images when pulled, and provide trust from publisher to consumer.

Docker File Security. Docker files build Docker containers. They should not write secrets, such as users and passwords. You should remove unnecessary setuid and setgid permissions, download packages securely using GPG and certificates, and try to restrict an image or container to one service.

Container Security. Docker security is about limiting and controlling the attack surface on the kernel. Don't run your applications as root in containers, and create specific users for testing and policing the Docker image. Run filesystems as read-only so attackers can not overwrite data or save malicious scripts to the image.

José provided a useful checklist to check the security of a Docker container, but it's not a short one. Remember, if you are going to deploy hundreds or thousands of these containers, you'll want to ensure consistent handling of security concerns to keep the hackers at bay:

- Do not write secrets to Docker files
- Create a user
- Follow version pinning for base images, packages, etc.
- Remove unnecessary setuid, setgid permissions
- Do not write any kind of update instructions alone in a Docker file

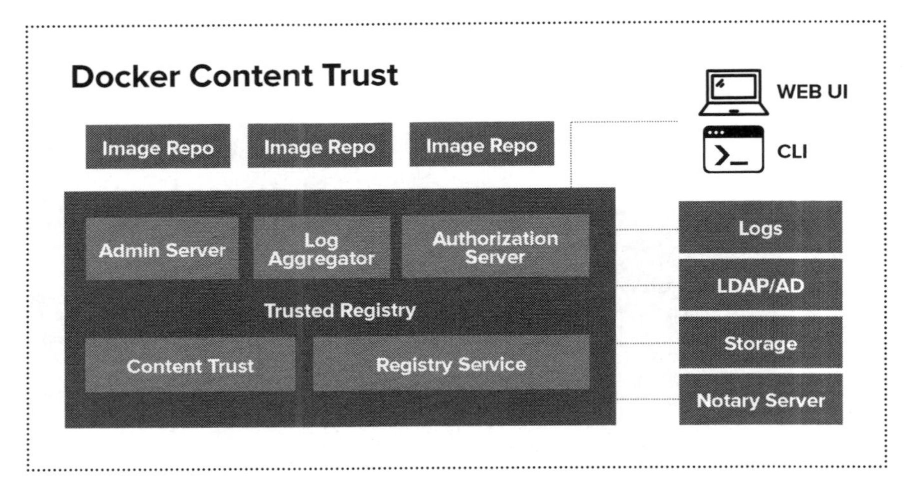

- Download packages securely
- Do not download unnecessary packages
- Use COPY instead of ADD
- Use the HEALTHCHECK command
- Use gosu instead of sudo whenever possible
- Use-no-cache (if applicable) when building
- Enable Docker Content Trust
- Ensure images are free from known vulnerabilities
- Ensure images are scanned frequently throughout your DevOps pipeline
- Ensure your images, packages are up-to-date
- Use file monitoring solutions for image layers (if required)

Auditing Docker Images. You can scan your images for known vulnerabilities with a wide variety of commercial and open source tools such as:

- Docker's native Security Scanning
- Sonatype's Nexus Lifecycle
- Twistlock,
- Docker Bench Security
- CoreOS Clair
- Dagda
- Aqua
- Tenable
- Anchore

All of these solutions can be integrated in one element of your CI/CD pipelines — some can be integrated in multiple places. In any case, you are well on your way to better image security with Docker.

CHAPTER 19

Docker: The New Ordinary

presented by Daniël van Gils

CHAPTER 19
Docker: The New Ordinary

The "new ordinary."

There you were. You heard about Docker, become mesmerized, floated into it's new special world...and then all chaos broke out. Things were not that easy and you scrambled for knowledge. Some folks saw the light and stepped forward, while others are still trying to figure the out the happy path home where containers will become part of their new ordinary.

This is the true story of Daniël van Gils (@foldingbeauty), a developer at Cloud 66, and the 12 steps he laid out in his Docker journey. His journey will take you full-circle, from your "ordinary IT world" to the "very special container world" and back to where the container world is the new ordinary.

Containerization continues to be a hot topic in the software world, and one that is only going to grow and mature. As it matures, questions still abound:

- How can you implement containerization into your organization?
- Have you already started but run into roadblocks?
- How can you be the hero of your organization's journey?

Daniël's journey started in the **Ordinary World (#1)** — the world you might be in now. You are at a conference, you read a book, you talk to some colleagues, and you want to jump into containers. You receive the **Call to Adventure (#2)**... but wait, you are human, you like the safe, you fear the unknown — so you want to **Refuse the Call (#3).**

Thankfully, others — like Daniël — have taken the journey. They can take you under their wings, coddle you, and then push you out of the nest. You **Meet a Mentor (#4)** and follow your heart. A heart is the good stuff — it is the core. Like most people would have, Daniël chose an artichoke to illustrate his point.

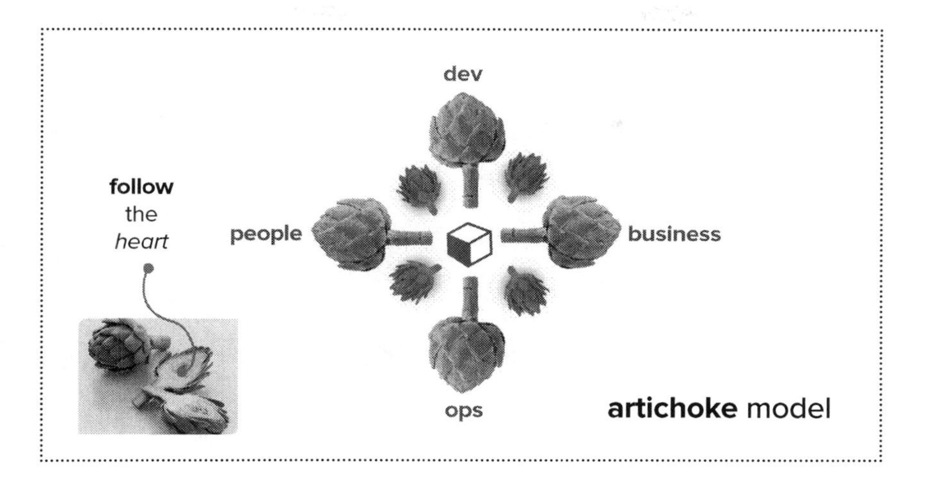

The heart of the artichoke is the good stuff, but the outer leaves protect the heart and make up the whole artichoke. Each aspect of an organization has something that drives them...their heart:

- Development — building elegantly performing code
- Operations — stable, performing code
- Business — delivering the right services to customers
- Customers — consuming the service they want

Your drive to satisfy the heart's desire gives you the courage to **Cross the Threshold (#5)**. You have a containerization machine, such as Docker. You put code in, creating nice little containers.

Now that your infrastructure is setup, you **test and know where your allies and enemies are (#6)**. The most important lesson here is that shortcuts are your enemy. Shortcuts are evil, promising a life of ease, but delivering a life of pain.

Daniël offers his **Approach (#7)** through the Docker journey to ensure you're successful. You must KISSSSS:

Keep It...

- **Slim** — remove what you don't need
- **Secure** — ensure you have the latest updates; remove all secrets

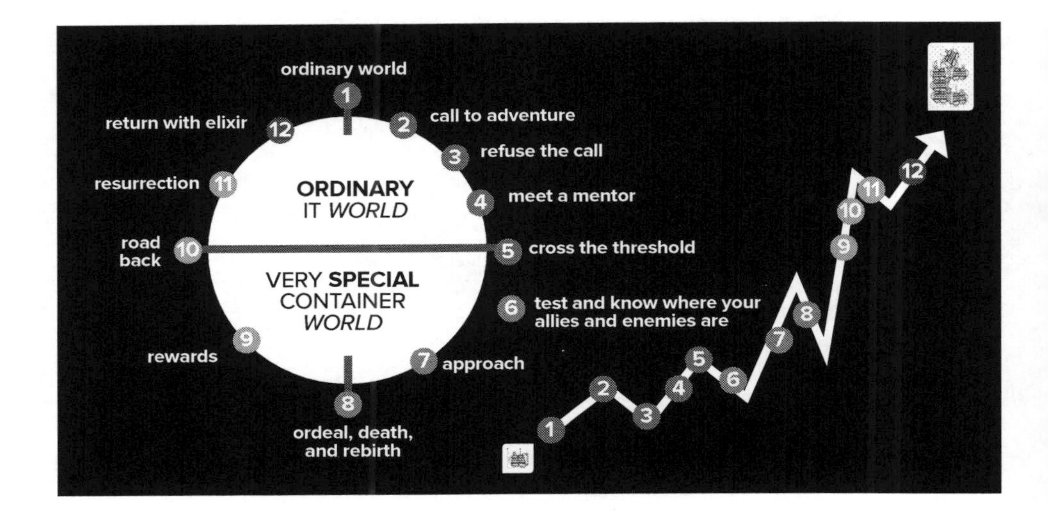

- **Speedy** — Follow best deployment practices and run performance tests
- **Stable** — Use version numbers for your Docker files
- **Set** — Immutable. Don't put databases or complicated volumes into your containers. You can, but the the technology is still too new.

The next steps in his journey cover **Ordeals, Death, and Rebirth (#8).** Once you pass this point in the journey, you'll understand what it takes to run Docker in production and have shown the proof of concepts. You'll then reap the **Rewards (#9)** and have clear visibility to your **Road Back (#10)** to the new ordinary.

For the successful heroes on this journey, now is your moment to shine. The next steps: your **Resurrection (#11)** and to **Return with Elixir (#12)** where containers are now the new ordinary.

Welcome to your new ordinary.

CHAPTER 20

Tickets Make Ops Unnecessarily Miserable: The Journey to Self Service

presented by Damon Edwards

CHAPTER 20
Tickets Make Ops Unnecessarily Miserable: The Journey to Self Service

From where I sit in the DevOps community, there is often more focus on dev than on ops. Damon Edwards (@damonedwards) of SimplifyOps seeks to change that.

Listen to Damon talk about DevOps best practices, and he dives right into the primary, systemic force behind most DevOps problems — silos. The product development process goes like this: Planning-->Dev-->Release-->Operate. The problem is the tendency in many enterprises to place similar functions together. Everyone ends up in a silo. Then walls build up between the silos. Eventually, people only know life in their silo, making handoffs even harder.

We often find application knowledge and business intent are heavily emphasized on development side but light on the operations side. Likewise, operational knowledge is heavy on the ops side, but light on the development side. Furthermore, development has ownership but limited accountability, while operations has accountability but no ownership.

While many enterprises are striving towards building cross-functional teams, the reality is that the transformations often stop short of truly integrating operations. The result? Siloes remain.

So, one has to ask, why is this so hard?

The reality is that enterprise operations are under tremendous pressure. One side is telling them to go faster and open it up and the other side is telling them to be more secure and be more reliable. These are often seen as competing priorities.

To solve this, enterprises need to "shift left" in the product development cycle operations activity as much as possible. They need to do as much as possible during development.

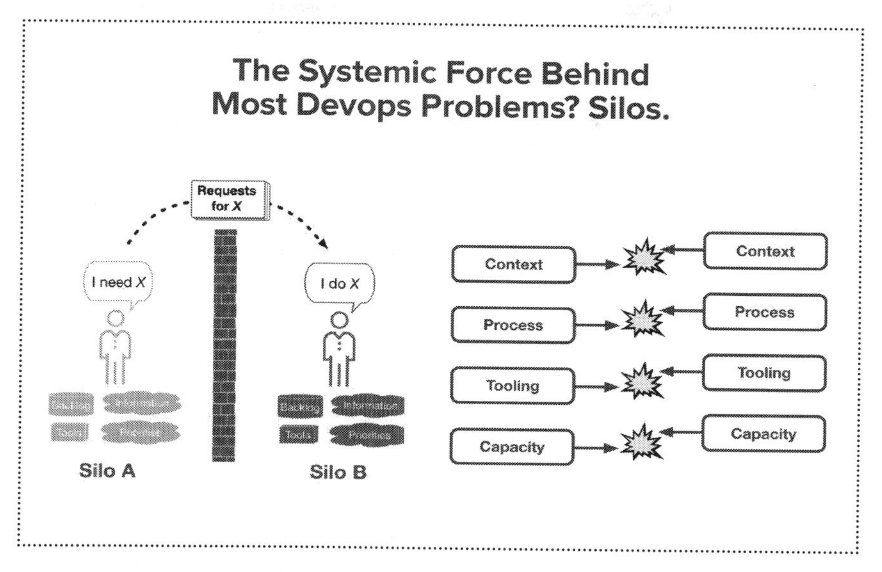

For the deploy function, enterprises should be:

- Writing/running automated tests
- Writing/exercising deploy automation
- Running security scanning tools

For the operate function, enterprises should be:

- Writing/exercising automated runbooks
- Writing/exercising monitoring/metrics
- Operational control (safely!)

However, shifting operations to the left is much harder. How do you do it? Embed ownership.

Let me repeat: *embed ownership.*

First, those who build something define the procedures to fix it, and those who build something fix it when it breaks.

That sounds simple, but raise questions:

- How do you safely and securely give out access?
- How do you enable the experts to contribute remediations?

- How do you give the experts visibility into operations?
- How do you do postmortems days/weeks/months later?

Damon recommended four steps.

Step 1: Establish a Secure Ops Portal

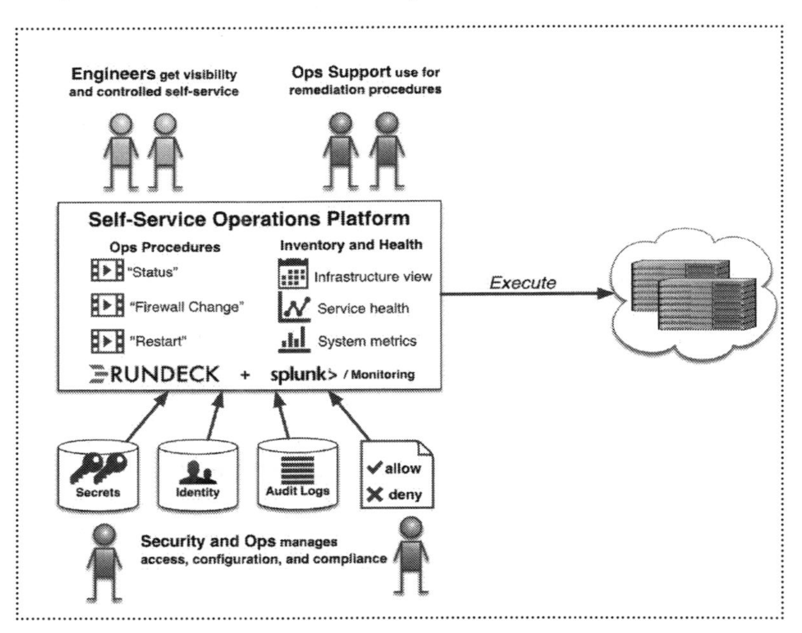

Step 2: Establish a SDLC for Ops Procedures

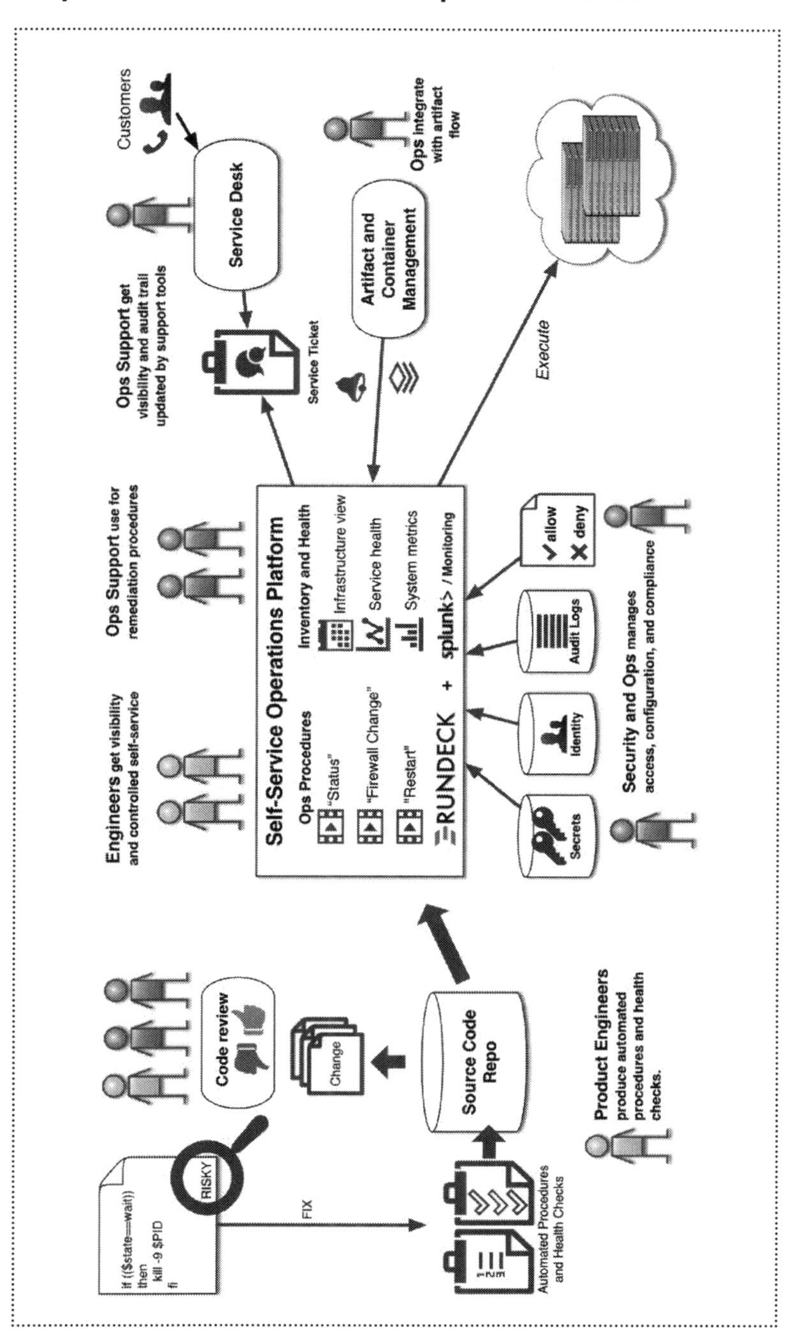

Step 3: Connect with Enterprise Management Systems

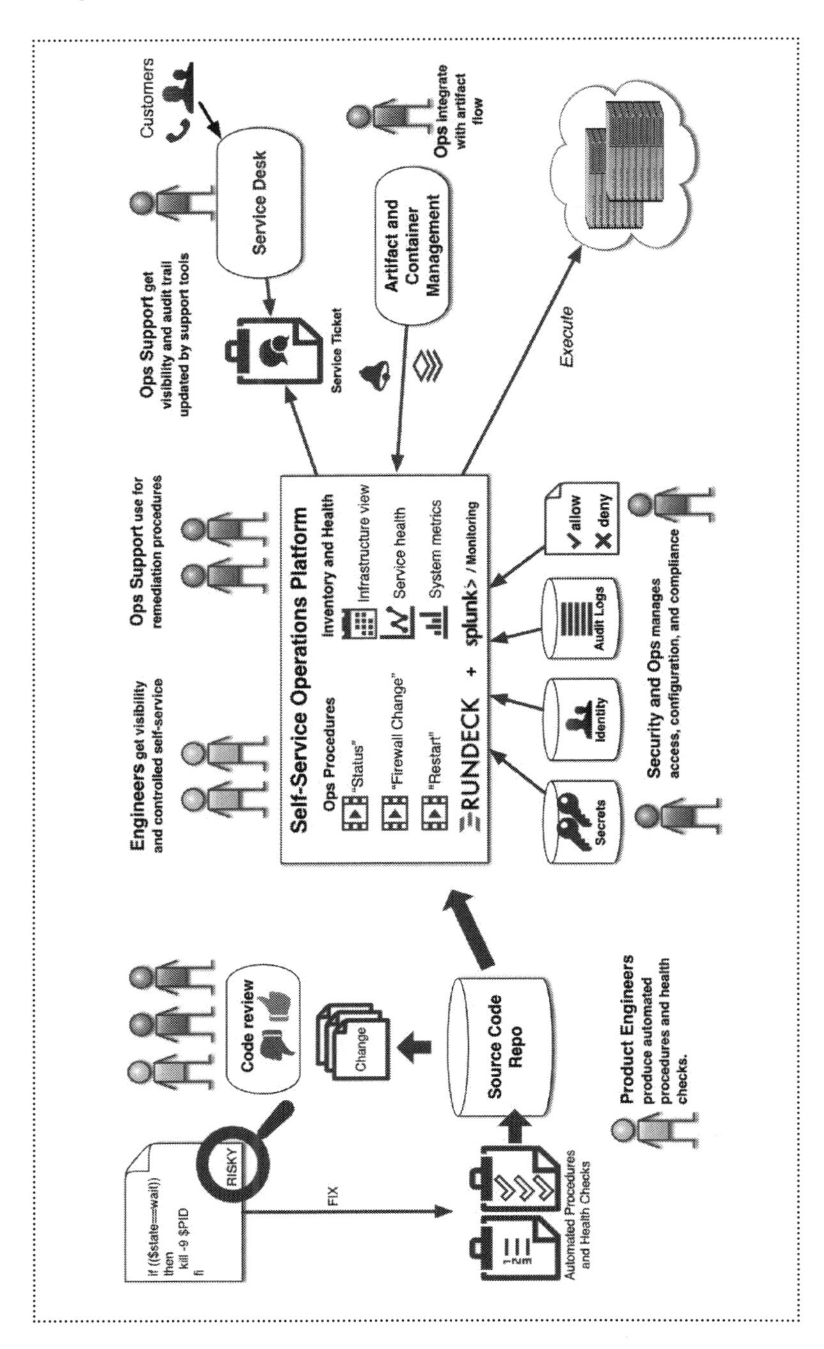

Step 4: Make Compliance Really Happy

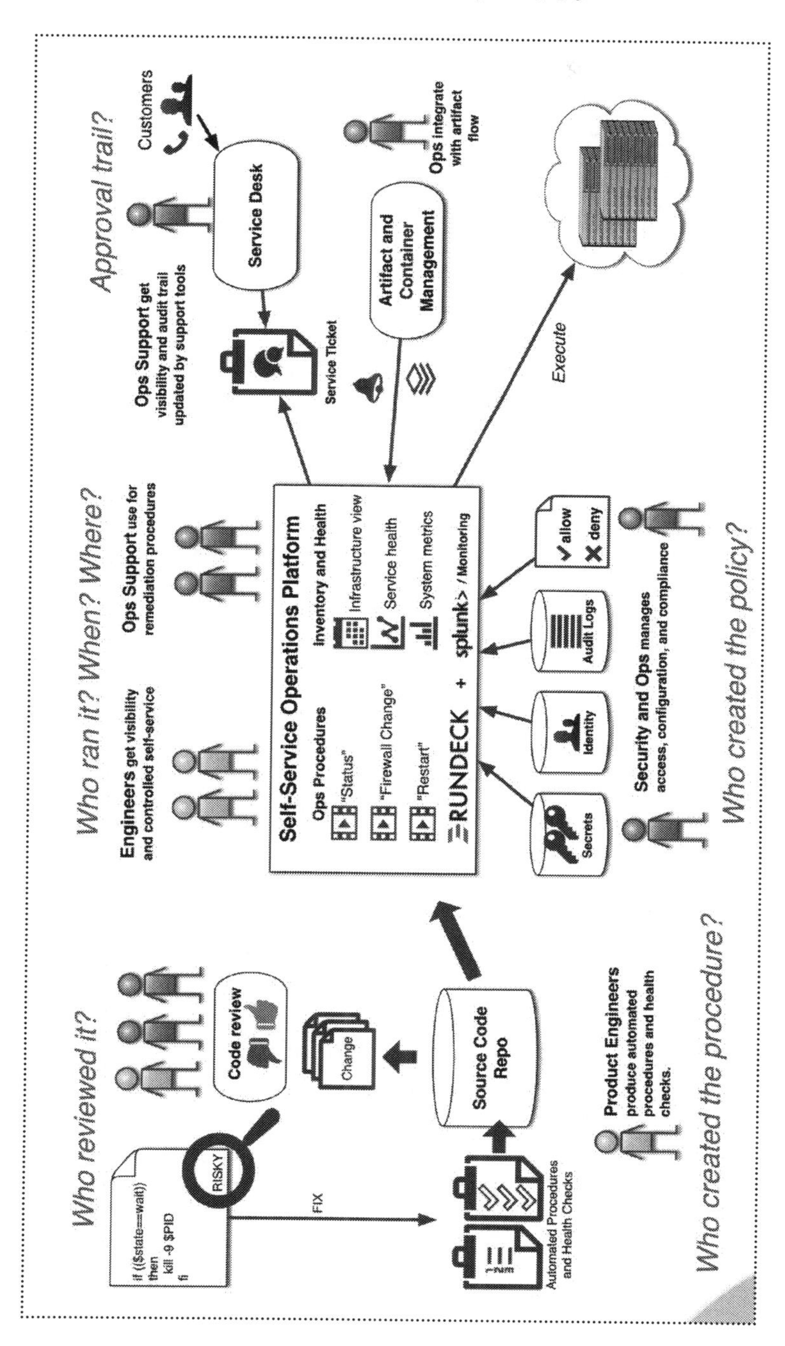

Ticketmaster is a real-life of example of this working at a large, prominent scale. Ticketmaster calls their system, "Support at the Edge" and it involves:

- Automated Ops procedures written/vetted by the delivery teams
- Ops remained in full control of what can run and security policy
- Empowered support teams with self-service ops tasks
- Empowered developers with limited self-service operations
- Combined with new incident response model

Ticketmaster has seen transformative results. Before Support at the Edge, the average mean time to respond was 47 minutes. Now, support at the Edge has reduced that to just 3.8 minutes in addition to decreasing escalations 50% and overall support costs 55%. Ticketmaster has seen real results.

While there is a lot to absorb from Damon's talk, if you just have one takeaway, it is to shift as much as possible left — into development.

CHAPTER 21

From "Water-Scrum-Fall" to DevSecOps

presented by Hasan Yasar

CHAPTER 21
From "Water-Scrum-Fall" to DevSecOps

As organizations abandon the waterfall method of software development for Agile, many are stuck in what Hasan Yasar terms Water-Scrum-Fall. That is, the organization has not effectively embraced Agile and DevOps principles and remains in silos with no links to business goals. Enter DevOps, an extension of Agile thinking. While Agile embraces constant change and embeds the customer into the process, DevOps embraces constant testing and delivery and embeds operations into the team to internalize expertise on deployment and maintenance.

Hasan lays out a plan to get organizations to DevSecOps. Really, DevOps is a risk mitigation strategy, built on situational awareness, automation, and repetition. But, security is where a lot of DevOps implementations fall down. The goals for each organization should be:

- Protecting private user data
- Restricting access to data/systems
- Protecting company data/intellectual property
- Standards compliance
- Safeguarding disposition/transition

But, how do organizations get there?

First, integration and communication. Every point of the product development life cycle should be integrated and communicating, including among the tools. Once this is achieved, you can automate many, if not most, of the tasks. The automated steps are the ones that require less human actions/input to the software development process. This allows everyone to focus on innovation and better code and less on tasks that can be automated by autonomous systems. Also, tasks that can be automated are less susceptible to errors.

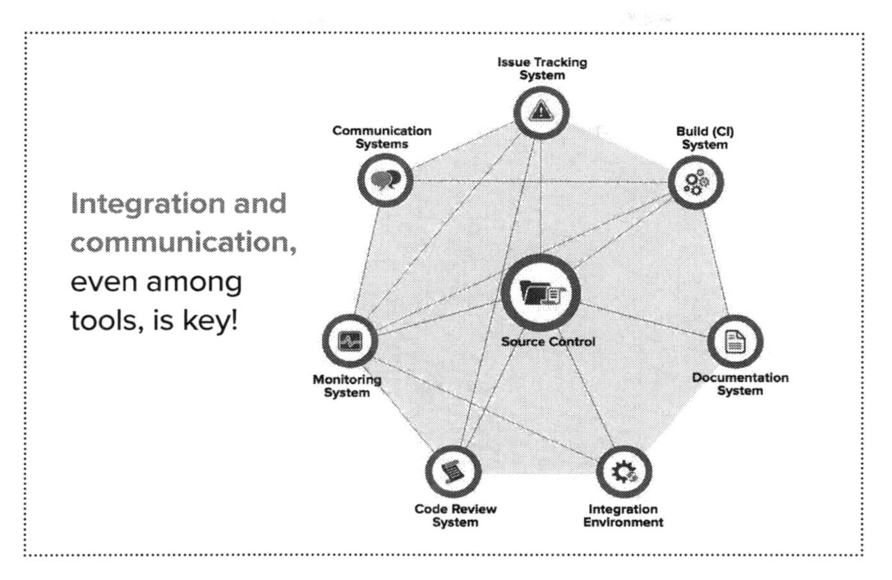

Of course, it is the team that ultimately designs, develops and delivers the software. Your team consists of development, IT operations, quality assurance and security. Each has its own skill set and focus, and the overlap is Secure DevOps.

The team is in place, processes are automated and development has started. Development in this day and age has evolved tremendously from even just a few years ago. Previously, software was limited to size, function and audience, and the supply chain was practically non-existent. Your team built each component. Now, development has grown beyond the ability of an organization to develop outside of its core competencies. The supply chain now involves many sources for the code. It is more like plug and play, and this creates lots of vulnerabilities.

Hasan notes the software supply chain risk factors:

- **Supplier capability** – Does the supplier follows practices that reduce supply chain risks?
- **Product security** – Is the delivered or updated product acceptably secure?
- **Product distribution** – Does the method of transmitting the product to the purchaser guard against tampering?
- **Operational product control** – Is the product used in a secure manner?

To reduce your supply chain risk, Hasan recommends:

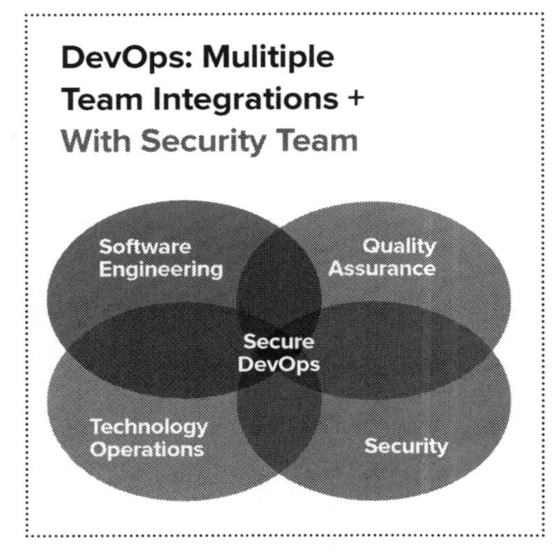

- Ensure supplier security commitment
- Evaluate a product's threat resistance
- Create a centralized private repository of vetted third-party components for all developers
- Establish good product distribution practices
- Minimize variation of components to make things easier.

Finally, as you transition to DevSecOps, remember that security must be addressed without breaking the rapid delivery, continuous feedback model. Otherwise, what is the point?

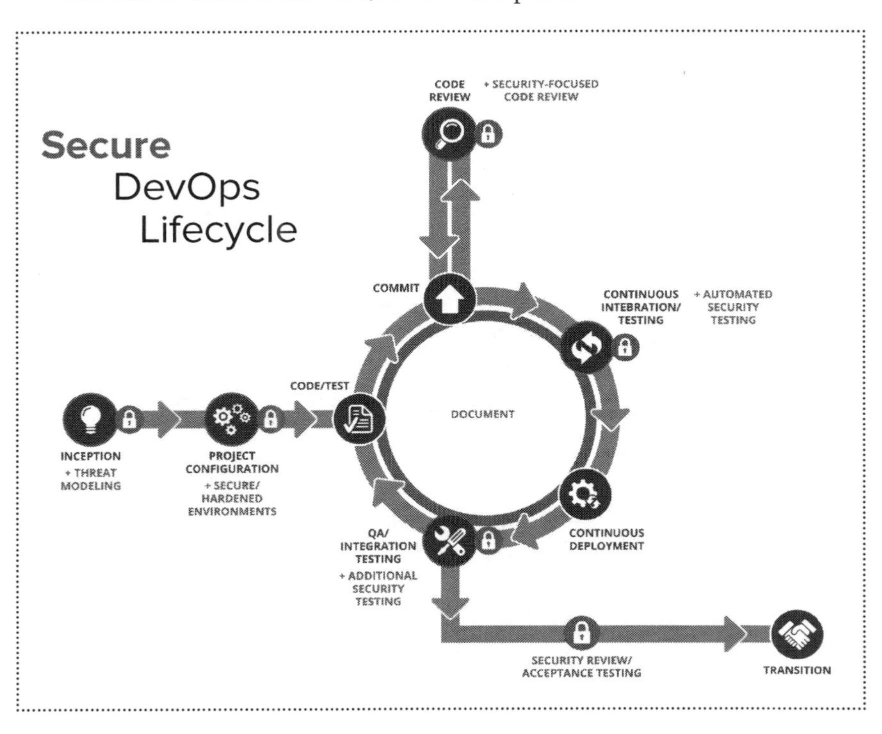

CHAPTER 22

How Capital One Automates Automation Tools

presented by George Parris

CHAPTER 22
How Capital One Automates Automation Tools

Listening to him talk, it seems like George Parris and his team at Capital One aren't keeping "banker's hours." George is a Master Software Engineer, Retail Bank DevOps at Capital One. George describes how they automated the DevOps pipeline for their online account opening project for Capital One, a major bank in the United States. Of course, there is a lot to learn from their experience.

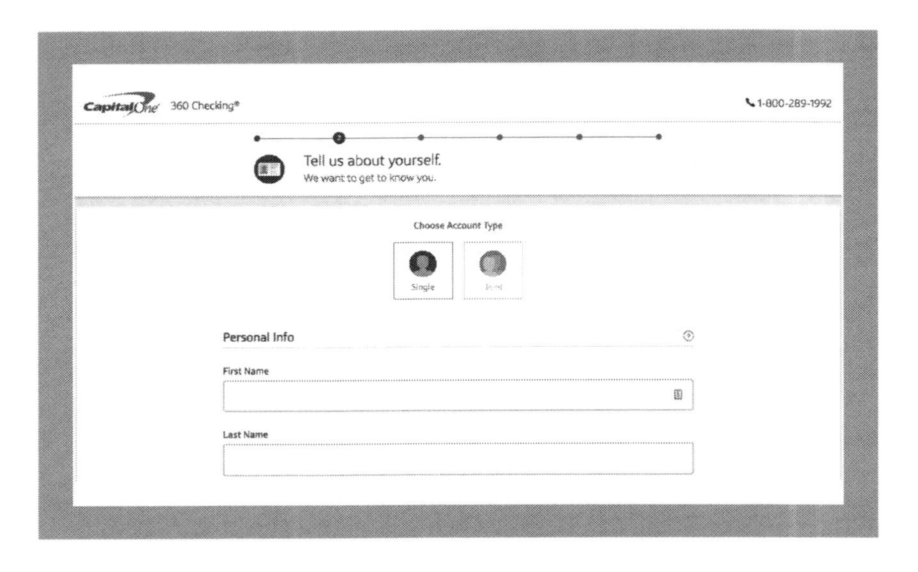

George started by pointing out that software development has evolved — coming a long way even in just the last few years. Developers now design, build, test, and deploy, and they no longer build out physical infrastructure — they live in the cloud. Waterfall development is rapidly being replaced by Agile, infrastructure as code, and DevOps practices.

How Capital One OAO deploys code:

Infrastructure as Code

- AWS
- Configuration Management
- TEST! TEST! TEST!

*IMMUTABILITY

Continuous Integration

Continuous Delivery... Approved and Scheduled Deployment

Capital One

Where we see these technologies and methodologies implemented, IT Operations teams are acting more like developers, designing how we launch our applications. At the same time, development teams are more responsible for uptime, performance, and usability. And, operations and development work within the same tribe.

George used the Capital One Online Account Opening project to discuss how they automate their automation tools — now a standard practices within their implementation methodology.

For starters, George discussed how Capital One deploys code (hint: they aren't building new data centers). They are primarily on AWS, they use configuration management systems to install and run their applications, and they, "TEST, TEST, TEST, at all levels." Pervasive throughout the system is immutability — that is, once created, the state of an object cannot change. As an example, if you need new server configurations, you create a new server and test it outside of production first.

They use the continuous integration/continuous delivery model, so anyone working on code can contribute to the repositories that, in turn, initiate testing. Deployments are moved away from the scheduled release pattern. George noted that, because they are a bank, regulations prevent their developers from initiating a production change. They use APIs with the product owners to automatically create tickets, and then product owners accept tickets, making the change in the production code. While this won't apply to most environments, he brought it up to demonstrate how you can implement continuous delivery within these rules.

Within all of this is the importance of automation. George outlined their four basic principles of automation and the key aspects of each:

Principle #1 — Infrastructure as Code. They use AWS for hosting and everything is in a Cloud Formation Template, which is a way to describe your infrastructure using code. AWS now allows you to use CFTs to pass variable between stacks. Using code, every change can be tested first, and they can easily spin-up environments.

Principle #2 — Configuration as Code. This is also known as configuration management systems (they use Chef and Ansible). There are no central servers, changes are version controlled, and they use "innersourcing" for changes. For instance, if someone needs a change to a plugin, they can branch, update, and create a pull request.

Principle #3 — Immutability. Not allowing changes to servers once deployed prevents "special snowflakes" and regressions. Any changes are made in code and traverse a testing pipeline and code review before being deployed. This avoids what we all have experienced — the server that someone, who is no longer around, set up and tweaked differently than anything else and didn't document what was done.

Principle #4 — Backup and Restore Strategy. A backup is only as good as your restore strategy. You know the rest.

And, with that, go and automate your automation.

CHAPTER 23

How To Fully Automate CI/CD — Even Secrets

presented by Andrey Utis

CHAPTER 23
How To Fully Automate CI/CD — Even Secrets

Imagine a world where your Continuous Integration/Continuous Deployment environment is 100% automated, including the passing of credentials.

Capital One, a leading U.S. bank, has achieved this, and Andrey Utis, Director, Software Engineering, at Capital One, outlined how they achieved.

Andrey started by stating what we all know — security needs to be at the forefront. Additionally, a message we hear over and over in DevSecOps is to automate where you can automate. That combination can be daunting and seemingly impossible. How can we automate security when someone needs to enter credentials and we can't store credentials where everyone can get to them? After all, credentials give you access to databases that often contain personally identifiable information and other protected information. A breach could devastate companies and people, whether intentional or accidental.

Andrey's team uses Amazon Web Services (AWS), so they give instances IAM roles that allow them access to other AWS resources. What is the key (pun intended)? On an AWS EC2 instance, there is a magic IP address to which you can make an HTTP call and it will return temporary AWS keys. Those keys then make the API calls to different database services.

Here is the solution. AWS KMS is encryption as a service. KMS Context allows you to add "salt" to the encryption. You can only decrypt with the same "salt." KMS Key Policies restricts which IAM role can decrypt with the key/salt. Master KMS keys can be used only to decrypt their own keys. Following is a code sample of a KMS policy.

```
{
  "Sid": "Allow use of the key",
  "Effect": "Allow",
  "Principal": {
   "AWS": "arn:aws:iam::123412341234:role/MyIAM Role"
  },
  "Action": [
   "kms:Decrypt",
   "kms:DescribeKey"
  ],
  "Resource": "*",
  "Condition": {
   "StringEquals": {
    "kms:EncryptionContext:AppName": "MyContext"
   }
  }
}
```

This allows multiple applications to be on one AWS account while limiting access of developers to the applications they are authorized for.

Here is the actual protocol:

1. Create a KMS master key (can be shared by multiple apps)
2. Create an IAM role for your server
3. Add KMS key policy that allows decrypt to your IAM role for a specific context
4. Encrypt your secret with the key and context and store the value in GitHub
5. At deployment or runtime, invoke KMS API to decrypt

There is a broader issue with IAM roles because credentials are generated by calling the "magic" metadata IP address 169.254.169.254. Developers in production, even with "read only" access to the instance, could call the KMS API to decrypt the secret. Developers should not be able to generate production IAM credentials at all, so you block the magic IP address with this code:

```
iptables -F INPUT
iptables -A OUTPUT -d 169.254.169.254 -m owner --gid-owner whitelist_group -j ACCEPT
iptables -A OUTPUT -d 169.254.169.254 -j DROP
service iptables save
```

To automate it and make it reusable, they created a Chef cookbook, which they call a "briefcase," to abstract decryption of secrets. They also have the iptables cookbook to block the metadata IP address for all except a whitelist of user groups, such as root and any application specific group that makes AWS API calls (see above code).

Andrey mentioned that Vault may have a viable solution soon. They use signed EC2 Identity Document to verify the caller. The current downsides are that it only supports authorization by AMI ID, but should support more soon, and secrets are not source controlled/versioned. As Andrey noted, that is not ideal for "configuration secrets" such as database passwords.

In the end, Andrey stated that the most critical lesson learned was that, "This is a new field. Most companies either don't fully automate or don't fully secure the entire pipeline. So, there is little information out there."

So, don't be the norm. Be different.

CHAPTER 24

Increasing the Dependability of DevOps Processes

presented by Ingo Weber

CHAPTER 24

Increasing the Dependability of DevOps Processes

For many users, software often isn't really appreciated until you don't have it. In this day, its constant availability has become a given, but, of course, 100% availability isn't really a reality. That is why when high-profile systems, like Netflix or AWS, have outages, it makes national news. Most of us don't work on systems that garner national user bases, but our users are just as important. So, we work hard to reduce system outages.

Where do problems arise that cause system outages? What can be done to improve processes to reduce system outages?

Researchers see that system outages often stem from problems during operations processes, such as upgrading software. Dr. Ingo Weber (@ingomweber) is one of those researchers. He is a principal research scientist and team leader at Data 61, a part of CSIRO Australia's government-funded research body. He and his fellow researchers developed an approach and tool framework, Process-Oriented Dependability (POD), to address this challenge in DevOps practices. POD enables fast error detection, root cause analysis, and recovery.

Ingo shared his insights on POD, describing the approach, tool, and some key findings.

Ingo set the stage by quoting a Gartner study showing that, "80% of outages impacting mission-critical services will be caused by people and process issues." Thus, showing that by addressing process issues, you can significantly reduce system outages.

He also notes that with significantly shorter release cycles, moving from months between releases and scheduled downtime to continuous delivery and releases delivered in hours or days, magnifies the potential issues. As an example, he notes that Etsy has an average of

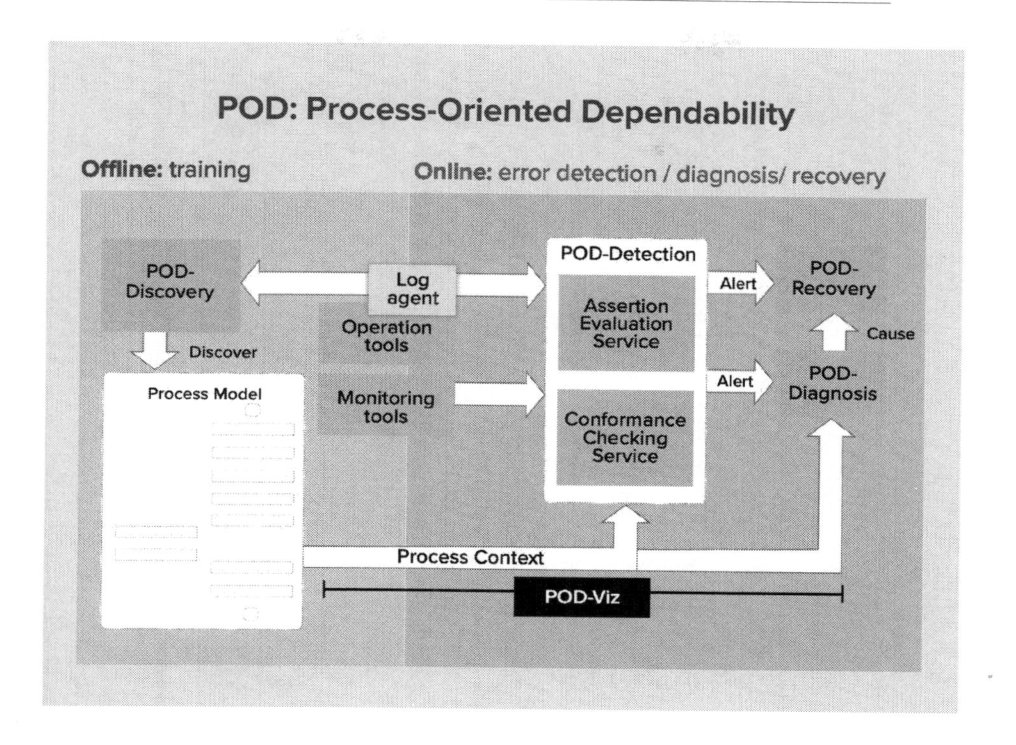

25 full deployments/day and 10 commits per deployment. Because of this, baseline-based anomaly detection no longer works because of cloud uncertainty and continuous changes, such as multiple sporadic operations at all times, scaling in/out, snapshots, migrations, reconfigurations, rolling upgrades, and cron-jobs.

The POD approach at a high-level is:

1. Increase dependability during Operation time through:
   » More accurate performance monitoring
   » Faster error detection
   » Fast or autonomous healing (quick fix)
   » Root cause diagnosis to figure out what the actual problem is
   » Guided or autonomous recovery
2. Incorporating change-related knowledge into system management

Digging a little deeper into POD, Ingo talks about two approaches they use: **Conformance Checking and Assertion Evaluation.**

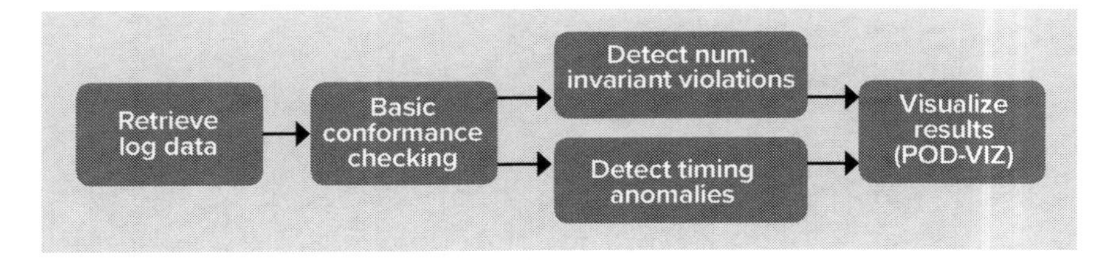

There are three levels of Conformance Checking:

- Basic
- Detecting numerical invariants
- Detecting timing anomalies

When errors/anomalies are detected, an alert is raised, and all results are visualized through POD-Viz, the dashboard.

Conformance Checking can detect the following types of errors:

- Unknown/error log line: a log line that corresponds to a known error, or is simply unknown
- Unfit: a log line corresponds to a known activity, but said activity should not happen in the current execution state of the process instance

All other log lines are deemed *fit*. The goal is 100% fit, otherwise raise an alert and learn from false alerts to improve classification and/ or the model.

Assertion Evaluation creates and checks against assertions. Assertions check if the actual state, at a given point, is the expected state. They are coded against cloud APIs so they can find out the true state of resources directly. You also identify the main factors affecting a resource and identify the log events that have the most important influence on changing the state of a system resource. Look at the metrics and chose whose are most relevant, then derive a formula that can be used to estimate the value of a variable associated with a system's resource so that you can test it against a range of acceptable values. Then, you drive an assertion based on this.

Does Process-Oriented Dependability sound like something you might want to implement or consider further? Ingo's suggested two papers: *Process-Oriented Dependability* and *Software Performance Engineering in the DevOps World*.

CHAPTER 25

Alert, Alert: Alert, Alert: Monitoring is More Than You Think

presented by Jason Hand

CHAPTER 25
Alert, Alert: Alert, Alert: Monitoring is More Than You Think

This is an alert about the importance of monitoring — and alerting. No, not the importance of monitoring and alerting to prevent incidents and outages. That is akin to telling a system administrator they need to do backups. This is about the unrealized benefits.

Jason Hand (@jasonhand), Senior Cloud Advocate at Microsoft, started by quoting a 2015 monitoring survey that asked, "Why are you collecting this data?" The survey reported:

- Performance analysis and trending 63%
- Fault and anomaly detection — 53%
- Capacity planning — 45%
- A/B Testing — 11%

What this indicates is what Jason calls the "tyranny of the service level agreement." That is that the first measurement of performance for customers is high availability for their servers (achieving "5 9s"), resulting in a prediction and prevention focus for operations. However, operations should be more than keeping the lights on and firefighting. Yes, you want to keep your customers happy campers, but customers want more than 99.999% uptime. They care about support, pricing, usability, and more.

Operations should be part of the larger systems thinking. Operations should be asking, "What are the business objectives? What does the business care about?" When you are just maintaining and firefighting, you don't have innovation? You need to be genuinely and generally learning and embrace the innovation that comes with it.

How important is learning and innovation? Blockbuster and Kodak are leading, recent examples of not learning and innovating.

Operations also shouldn't be all about prediction and prevention. The better metric for systems is Mean Time to Recover rather than Mean Time Between Failures. We can't engineer out failure, so we should learn how to address failure quickly. In complex environments, there are going to be problems, but when they are resilient, they have the ability to "resist, absorb, recover from or successfully adapt to adversity or changing conditions."

Monitoring and alerting give you information to help you learn, improve, and innovate. You can use the information about events in the past to drive future actions. You can find better ways to respond to problems and how to deliver and support your services.

However, you must have a culture of learning, and learning doesn't come from reading and listening. Learning comes from doing. Real learning comes from: observing → orienting → deciding → acting. You go from knowing to understanding to learning. All of this requires mistakes and a feedback loop and a culture that looks at improving rather than blaming, one that trusts operations is doing the best job given the constraints of the system.

Jason wrapped up his talk by underscoring that learning and innovating leads to uncovering new ways of building, deploying, and maintaining software and infrastructure, which leads to resilient systems. The by-product of a highly resilient system is a highly available system, and that the unrealized role of monitoring and alerting is learning and innovation.

CHAPTER 26

Leading a DevOps Team at a Fortune 100 Company

presented by Uldis Karlovs-Karlovskis

CHAPTER 26
Leading a DevOps Team at a Fortune 100 Company

A manager, an evangelist, and a godfather all walk into a bar.

Okay, this isn't a bad joke, this is an article about how managers can implement DevOps in large organizations, especially where culture and organizational change hamper efforts. But, I promise, we will talk about managers, evangelists, and godfathers.

Uldis Karlovs-Karlovskis (@UldisKK) works at Accenture Latvia. Accenture is a global consulting giant, employing 400,000 with teams and clients across the globe. Uldis helped bring DevOps into the organization and shared his journey — the good and the bad.

Uldis joined Accenture in 2012 and faced three top challenges:

1. How to maintain quality at such growth?
2. How to enable continuous learning?
3. How to stay fast and flexible at their size?

His team was small, they held daily stand-ups that were productive, and they were flexible. Life seemed good, and then his team grew into a "spider." They still held daily stand-ups, but they were useless because it was just people reporting and no one understood what is going on.

What did Uldis do? Well, he didn't freak out, if that is what you are wondering. Actually, he did. One morning he, "woke up, realized I had 25 people under me, freaked out, and skipped the job for the day." He walked around, thinking about it, and wondering what to do. He thought about his 3 to 4 types of DevOps projects and all of the non-project activities that were monopolizing his time. He realized he focused on the urgent over the important tasks, working 10-12 hours/day.

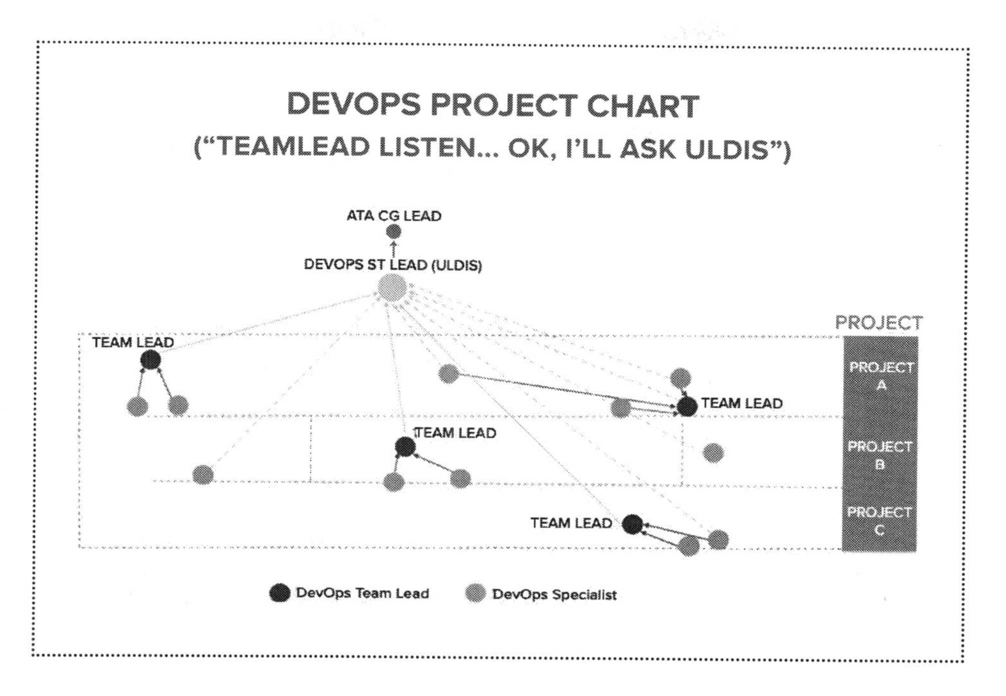

What is a manager to do? Have people help him, of course. Uldis decides to create an unofficial role that sounds interesting — a subject matter expert (SME). They would work on non-project activities, increase team engagement, improve satisfaction, and asses DevOps maturity in projects. Initially, people were motivated to be SMEs, but they were always busy with other things. So the effort failed.

Uldis had a new idea — instead of him leading the SME team, he will let them decide who wants to do what, they can join the SME team they want, each team can pick a lead SME, set their own goals, how they work, etc. This helped, but still didn't solve the problems. Uldis is still burdened with the tyranny of the urgent.

Back to the drawing board, and a new idea is born — evangelists. Uldis broke down 50 SMEs into 7 sub groups and added a little hierarchy because the flat structure wasn't working. Each evangelist lead a guild — one of seven. Team members were to talk to their evangelist before coming to Uldis with questions.

But, alas, Uldis was still just handling escalations and talks.

Evangelist Responsibilities

Expertise in specialization

Training roadmaps

Engagement of people
(sync with SME group motivators)

Regular recognition of people

Talent discussions, rewards, outcomes twice per year

Get input for talent discussions from projects

Intern-to-employee discussion

Consult on which career counselor to choose

So, what do you turn to in order to bring order — bring in godfathers.

Now, team members can always go to the Godfather for help. Uldis decides that, generally, each person has four topics they may want to talk about. For each topic, they have a person to talk to:

- SME addresses how to contribute to the company in general
- Godfather addresses escalations, such as not happy with the Team Lead
- Evangelist addresses questions about the person's skills
- Team Lead addresses project tasks

Uldis now realizes the benefit of this approach — in a way, now, everyone is focused at some level on sales. As we all know, this is very hard, but important, in any organization, but especially when everyone is an engineer.

While his approach has helped him and his team, he also acknowledges that, "If I did the same thing today, I would fail, but maybe for different issues."

So, he is looking at improving his culture. As we talk about at All-

Godfather Responsibilities

Discuss new projects

Help Solution Architects

Agree on future staffing plans

Keeping an eye on a project

Being a point of escalation, after Team Lead

Current staffing issues

Closing/ending a project

DayDevOps time and time again, good culture is key to successful DevOps. Uldis, looking at Accenture, notes they don't fit into a typical organizational structure. Reading literature, industry buzz, etc., he realized something was missing from these models, but he wasn't sure what. He does know they needed to be better.

He discusses three dominant types or organizational cultures:

- Pathological — 8 hours of work and your job is done
- Bureaucratic — 2-12 hours of work if the process requires it
- Generative — 2 to 12 hours of work and I'll deliver it

Uldis is trying to be at the third approach, and one key to it is if you have a bad day, leave it for tomorrow when you will be more productive. Tomorrow it may take more than 8 hours, but you will put your best into your work.

His journey is ongoing, but he finished his talk covering a few things they do on his team:

- Onboard new joiners through a continuous delivery pipeline
- Meet weekly lunch
- Half-yearly performance assessments

He also provided four takeaways:

- Everyone by default is good
- Intrinsic over extrinsic
- Engagement is your work responsibility
- Let them lead

Uldis told an honest and open story about his journey with the hope it will help others, because, while parts of it can be fun, it is no joke. It is hard, but worthwhile.

Half-Yearly Performance Assessments

1. 55% TOP-DOWN — Project Manager or Team Lead feedback
2. 15% BOTTOM-UP

    360 feedback for team
    Self-assessment for one-man army

3. 15% BIRD'S EYE — Career Counselor feedback
4. 15% EXTRA MILE — Evangelist feedback

CHAPTER 27

Microcosm: Your Gateway to a Secure DevOps Pipeline as Code

presented by Hasan Yasar

CHAPTER 27
Microcosm: Your Gateway to a Secure DevOps Pipeline as Code

Development pipeline: "an automated manifestation of your process for getting software from version control into the hands of your users."

Seems easy, right? Okay, not really. There are key questions to ask first. Who owns the integrated pipeline? What and how do you measure and monitor in order to assess pipeline health? What are the key qualities and attributes teams should look for? Oh, and there are 180 some odd tools available to fit in your DevOps pipeline.

To build your pipeline, you will need to assemble and integrate many moving parts. Of course, you'll also want it to work with the first real deployment. After all, you've spent a lot of capital convincing your organization this was a worthwhile investment, but they are still nervous and skeptical.

If only there was a way to see and understand a pipeline without the large, initial investment of resources and even more precious time.

Enter Microcosm

Hasan Yasar (@securelifecycle) brought this topic to life by discussing Microcosms. Hasan, who works at the Software Engineering Institute (SEI) at Carnegie Mellon, explained that Microcosm was developed at SEI as a miniature, secure DevOps pipeline that is available through infrastructure as code. It is a miniature version of what you would find in a large organization and is designed to help introduce people to development pipelines.

Stepping back a bit, Hasan reminds us that DevOps is about, "breaking down the communication silos to establish effortless efficiency/collaboration between teams because we're all on the same team, striving for the same goal!" A deployment pipeline helps achieve this goal by integrating security into the deployment process.

To start assessing your development pipeline, Hasan laid out a number of key quality attributes in order to select the right tools.

Key Quality Attributes of a Pipeline

- Integrate-ability
- Interoperability
- Usability
- Portability
- Resilience
- Security/Permissions
- Availability
- Scalability
- Performance
- Modifiability
- Configurability
- Automate-ability of manual tasks
- Approvability — allows for manual approvals
- Measurability
- Others, based on the project

These attributes will help you seamlessly inject security at multiple points into a development pipeline, illustrated on the following page.

But, what about Microcosm? Well, it consists of four virtual machines and creates a secure DevOps pipeline via IaC using Vagrant, Chef, and Ansible. Each of these services is integral, but, working together, they are invaluable and create a Continuous Integration and Continuous Deployment platform with Secure DevOps best practices.

The first virtual machine offers:

- Jenkins CI/CD service
- OWASP ZAP web application security scanner
- Selenium web application software-testing framework

The second virtual machine offers:

- GitLab repository manager

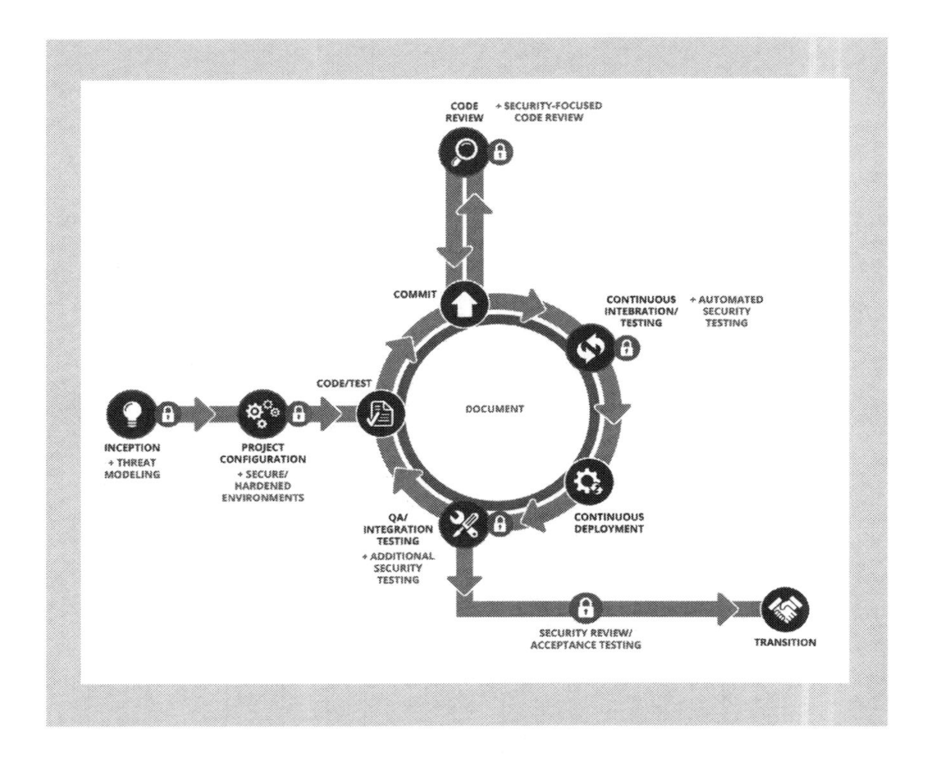

The third virtual machine offers:

- Media/Wiki service
- Bugzilla issue tracker
- Hubot chat bot

The fourth virtual machine:

- Acts as a staging server for deployed instances of PetClinic Spring web application

All services and the project and open source and you can clone the repository at https://github.com/SLS-ALL/devops-microcosm.

What is Next?

They plan to offer a microservice version and one with Docker containers/Docker Compose, and continue to update Chef recipes of services used to secure vulnerabilities.

CHAPTER 28

Modern Infrastructure Automation

presented by Nathen Harvey

CHAPTER 28
Modern Infrastructure Automation

"We're no longer an airline. We're a software company with wings," claims Veresh Sita, CIO of Alaska Airlines. The success of today's businesses rests on software. It is an integral part of infrastructure, so we need to always ask, "How can it be better? More reliable? More secure?"

Nathen Harvey (@nathenharvey), at the time of this presentation, was the VP of Community Development at Chef and the co-host of the Food Fight podcast. He is also an evangelist of automating software operations.

Nathen started his talk by stating that "automation at scale is required for high velocity IT," making the point that in today's environment, you can't wait months between deployments. Enterprises are moving to Continuous Integration/Continuous Deployment, and he advocates for Continuous Automation.

To implement automation at scale, you have to start with a dynamic infrastructure by:

- Provisioning and setting up environments
- Implementing dynamic scaling of compute resources
- Migrating legacy workloads to the cloud
- Deploy in multi-cloud and hybrid-cloud environments
- Support heterogeneous environments

Several speakers of All Day DevOps talked about automation, and all of them tell you a critical step is to treat infrastructure as code. According to Nathen, this means that to programmatically provision and configure components, such as servers, databases, firewalls, etc., you also have to treat it like any other code base. This includes version control and automated testing, but also:

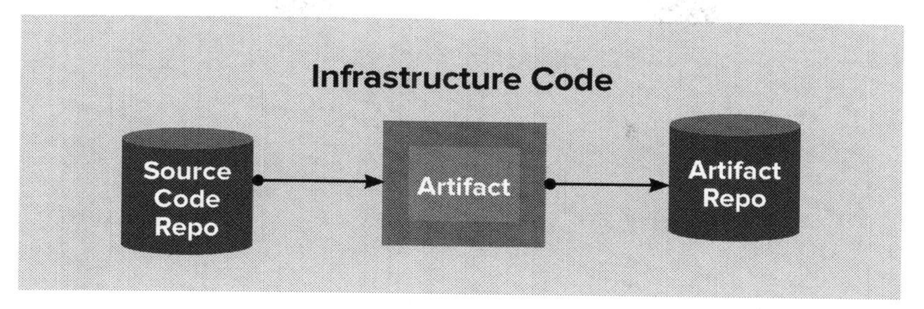

- Automated, full-stack application policies
- Package and service installation
- Versionable, testable, and repeatable workflow
- Scalable application policies
- Management of interdependencies across nodes

When you treat your infrastructure as code and deliver at scale, you can scale elegantly from one to tens of thousands of managed nodes across multiple complex environments.

In order for software to be effective, users have to trust that it will do what they expect it to do. This is true from handheld calculators to self-driving cars, and it is also true when we automate our development and operations. Nathen advocates that we have to build a pipeline to production that creates faith in us that our code is ready for production. This pipeline must:

1. Test the code/verify locally
2. Verify in pipeline with automated testing
3. Approve with another engineer (aka code review or Nathen's "four eye rule")
4. Build an artifact
5. Deploy into acceptance environment to automatically test the artifact
6. Ask someone else — do we want to ship this?
7. Ship the code with a click of a button
8. Union (bring together dependencies and run tests)
9. Rehearsal
10. Delivered (ship to customers)

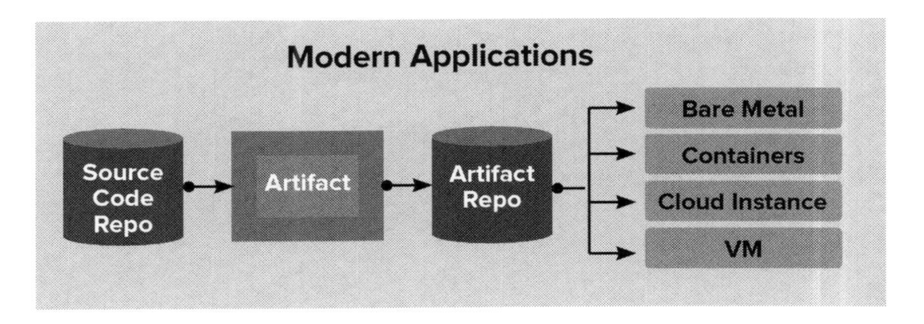

What Does the Future Hold?

Infrastructure automation is NOT enough. Why? Auditors show up to audit your infrastructure and applications. The problem is they use different tools — Excel spreadsheets, notebooks, PDFs, etc. This makes it very difficult to audit. Nathen advocates managing compliance documents as code, too. When we translate compliance into code, it can be executed as part of the pipeline so we can test for compliance throughout the entire lifecycle.

Finally, Nathen stresses the need to have application automation too. Modern applications look like infrastructure as code.

Developers can make applications that operations can deliver and run in production. This makes it easier to consume, run, and keep applications healthy. Operable applications need to be:

- Isolated
- Immutable
- Configurable
- Rebuildable from sources
- Built on a common interface for monitoring health
- Reside in common packaging
- Maintain runtime independence

As enterprises apply infrastructure, application, and compliance automation, it creates a freedom for people to focus, and people are the key.

Nathen notes that "DevOps is a cultural and professional movement, focused on how we build and operate high velocity organizations, born from the experiences of its practitioners." He underscores that people are the future of automation. Automation places the priority on people, which enables great companies and great products. So, be great. Automate.

CHAPTER 29

Nothing Static About the Growth of Static Analysis

presented by Justin Collins

CHAPTER 29
Nothing Static About the Growth of Static Analysis

100:10:1. The Ratio of Doom. For every 100 developers, there are 10 people in operations, and 1 person in security. It is called the Ratio of Doom because this can spell disaster for enterprises. One solution can be to automate security early and often in the development process with static analysis tools.

Justin Collins (@presidentbeef), is the creator and maintainer of Brakeman, a static analysis tool for Ruby on Rails applications.

CREDIT: SHANNON LIETZ

Justin notes that in DevOps developers are as responsible for stable code as the ops team is. In DevSecOps developers are as responsible for secure code as the security team is. Each has different roles, but they all have a responsibility. The security team's role is to provide:

- Expertise
- Guidance
- Training
- Tools

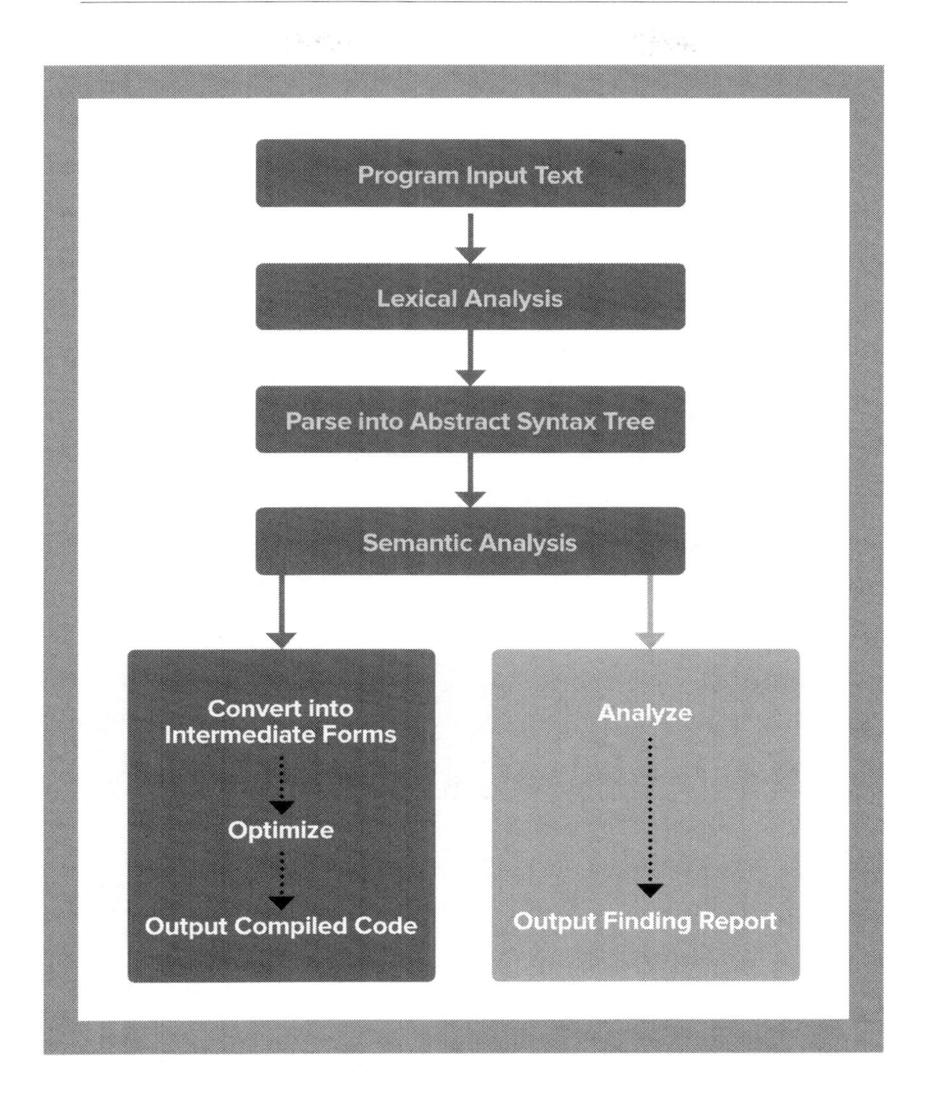

Tools are vital, and in DevOps tools need to be:

- Automation friendly
- Fast
- Consistent
- Providers of early feedback for developers

Of course, one set of tools are analysis tools. To help explain what they do, Justin used a compiler as a parallel to a static analysis tools, except that you get a report instead of executable code. As he said, "It isn't magic — it is just a compiler with a different target."

To see how static analysis tools measure up to Justin's recommendations for DevSecOps tools, he asks, "Is it automation friendly?" If it is source code analysis, then, yes, it is. As an example, Justin provided a "Tweetable" incremental scan for Brakeman.

Tweetable Incremental Scan

```
if [ -e "report.json" ]
then
  brakeman --compare report.json -o diff.json -o report.json -z
else
  brakeman -o report.json -z
fi
```

"Is it fast?" If you are comparing it to web scanners, then definitely yes, but many are not fast, yet. However, Justin predicts that with the momentum of DevOps, the tools are going to have to be fast. So, there will be next generation static analysis tools that will be fast and easy to automate.

"Are they consistent?" Yes. You need to have consistent results with a baseline scan and then incremental results for comparison. Static analysis tools deliver this, unlike web scanners.

"Do they provide early feedback for developers?" Security needs to shift left in the development process so that developers start thinking about security and writing it into their code.

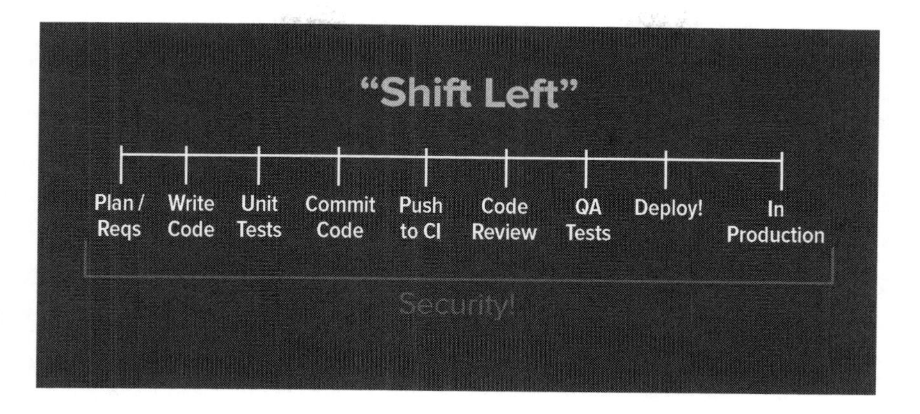

Early security feedback in the development process is important because:

- Fewer dependencies makes integration easy
- Fast tools can be "in line" with workflow
- Incremental results relevant to changes

The ideal place to start source code analysis is at "Run in IDE/On Save." Brakeman, for example, can run when the code is being saved. As Justin stated, you can really do it at any point in the process that you have code.

Many static analysis tools certainly meet the four requirements for security tools that Justin outlined. Of course, no solution is perfect, and the problem with static analysis is that every language needs it own tools, and it needs to understand the frameworks used.

There are also several types of static analysis tools:

- Security — vulnerabilities
- Composition — old/vulnerable dependencies — known public vulnerabilities in libraries
- Quality — complexity
- Code coverage
- Style

Thankfully, Justin provides some lists to help you find an appropriate tool for you: Awesome Static Analysis and OWASP.

Finally, Justin concluded with four points:

- Source code analysis fits well with DevOps
- Enables security review inside workflow
- Provides feedback early in development
- Has multiple options for integration points

So, no matter the tool, know that security automation is vital. Choose the tools that are right for you.

CHAPTER 30

One Team, 7,000 Jobs: Life in the DevOps Jungle

presented by Damien Coraboeuf

CHAPTER 30
One Team, 7,000 Jobs: Life in the DevOps Jungle

Damien Coraboeuf (@DamienCoraboeuf, nemerosa.com) has 7,000 jobs. While you might gasp at that workload, Damien is not stressing out. All 7,000 jobs are automated within his team's Jenkins pipelines. How does he do it? Damien follows four key principles to keep his cool in the job jungle: self-service, security, simplicity, and extensibility. But you might be surprised that one of his most important survival techniques is treating his pipeline as "not code."

"Look Ma, I've built a pipeline!"

"Lovely, now, build one for your sister."

Even with 7,000 jobs in Jenkins, Damien has a tiny team of six running it while supporting a large band of developers around the world.

Damien laid out how a pipeline, the backbone of a Jenkins job, is constructed. Spoiler alert — it is pretty simple, hence the magic of Jenkins. However, it gets complicated quickly, so Damien shared four principles his team strive towards:

- **Self-service.** They don't want the Jenkins team to be a bottleneck, and they don't want to have to deploy an army just to run Jenkins.
- **Security**. With teams of developers, many of whom they have never met, security and control is critical.

- **Simplicity.** Developers can focus on developing rather than Jenkins.
- **Extensibility.** As they grow and capabilities are dreamed up or implemented, their system needs to be able to adapt.

Core to their implementation and their goals of self-service, simplicity, and extensibility is the Job DSL Plugin. While Jenkins has a robust user experience, as jobs are added, it can get cumbersome and clumsy. The Job DSL Plugin brings order, automation, and consistency to the jobs. Damien walks through the basics of the plugin and its benefits, noting that it treats the pipeline as code.

Treating the pipeline as code allows the team to reuse code through versioned libraries and easily manage thousands of jobs. Requiring developers to manage more work as code does offer benefits, but security risks also come along with giving that many people access to the code. So, Damien's team actually treats the pipeline as "not code," by

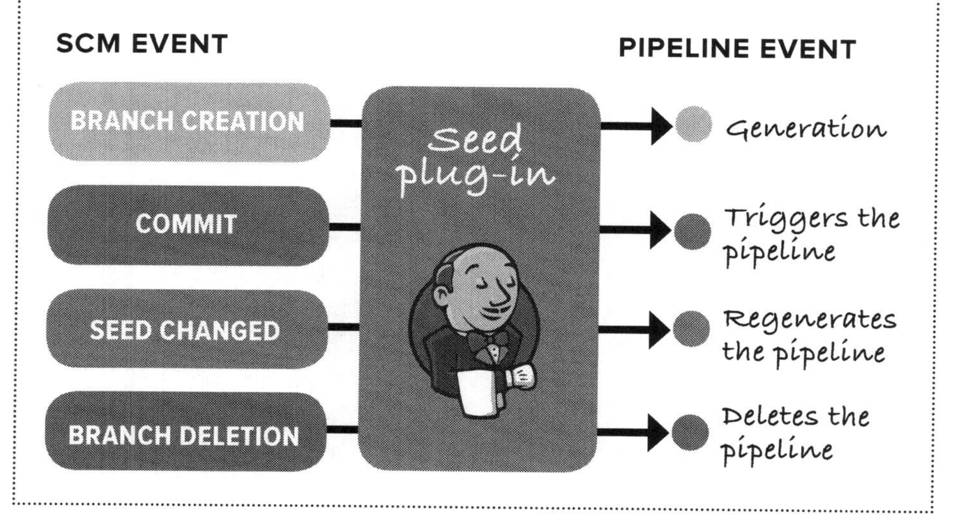

describing the pipeline using properties files. It also allows them to use the pipeline as metadata, so queries and reports are almost enjoyable.

Damien's team also utilizes the Seed plugin, which they maintain, to, "help automate the generation and management of pipelines," tighten security by keeping projects locked to teams, and increase automation through hooks.

The Seed plugin shares some functionality with the popular Pipeline plugin, which is maintained by a large community. However, the Pipeline plugin does not allow for treating the pipeline as "not code." Damien's vision is that one day Seed will be an extension of Pipeline.

Damien's team has done a tremendous job of running Jenkins across the enterprise and demonstrates what can be done with automation

CHAPTER 31

Organically DevOps: Building Quality and Security into the Software Supply Chain at Liberty Mutual

presented by Eddie Webb

CHAPTER 31

Organically DevOps: Building Quality and Security into the Software Supply Chain at Liberty Mutual

Some people are directors, managers, and administrators. Others are disruptors.

Eddie Webb (@edwardawebb) iis an IT Disrupter for Software Development Platforms, formerly with Liberty Mutual. Eddie's talk looked at Liberty Mutual's transformation to Continuous Integration, Continuous Delivery, and DevOps. For a large, heavily regulated industry, this task can not only be daunting, but viewed by many as impossible. He noted that DevOps is not their goal; Continuous Delivery is their goal, and DevOps is the enabler.

It all started at Liberty Mutual in 2004 with a vision: "Black-box reusable services that benefit the business." They also had a concept that "developers own the quality of the code; we have no QA team." Employees were empowered with this "concrete vision but malleable strategy," as Eddie said. They gave them time and space to explore solutions, because ownership increases adoption.

The team started with fixing what frustrated them, such as: CI/CD pipelines, artifact and code repositories, elastic cloud-native runtime, and Agile planning and delivery.

But what about centralizing DevOps?

The reality is that tools can be centralized, but culture can not.

Eddie laid out an example of a large organization's IT structure, saying that this is how things work at Liberty Mutual. Eddie's team is the Central IT, and their customers are the Market IT team, not Liberty Mutual's end customers. As Eddie pointed out, the more layers you create, the more friction that enters the system. Often, organizations try to reduce the friction through micro-fixes, but Eddie's team asked how to change the culture to reduce the friction.

This required distributing innovation and centralizing the tools that work. As an example, if you build a manufacturing plant, you wouldn't build your own power plant. If you are in IT, why build your own cloud?

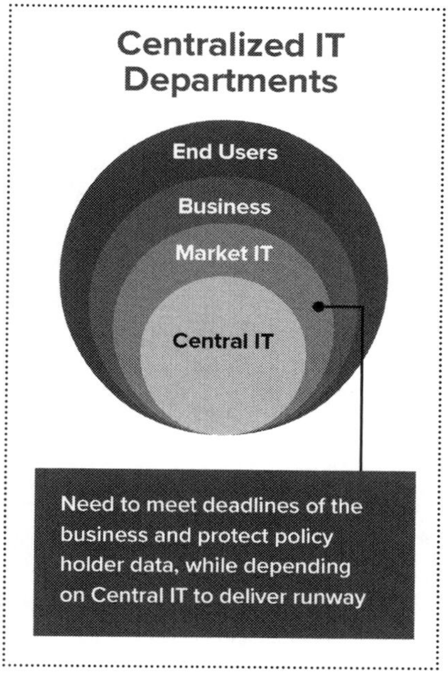

This is the Netflix way. At Netflix, teams are independent, but responsible. Engineers can choose the best tools for the job. No one single team is responsible for innovation. However, the Engineering Tools team helps direct other teams' experimentation toward new products without getting in the way of innovation.

Eddie quoted *Drive* when looking at how to motivate innovation in complex work environments. The author says if you want to incentivize or motivate cognitive tasks, you need three things: autonomy, mastery, and purpose.

When you centralize non-value added platforms, it enables teams to:

- Master their specific domains of business
- See clear purpose in everything they build
- Maintain autonomy of choice of frameworks, patterns, and approach

When you decentralize, it raises the question — what about security? Developers are humans, and humans inherently try to avoid tasks they see as unnecessary or overly burdensome, and this becomes easier when the compliance person isn't there (think of how we handle speed limits). Traditional security measures erode profit and market share, and, from a stat Eddie quoted from Gartner, 77% of IT security professionals agree that security slows down IT.

Whether you are centralized or decentralized, security (and other compliance-oriented tasks) can increase adoption rates by making it easy to do the right thing. How do you do this? Eddie proposes governing with tools (think of speed governors in fleet cars).

You also need to have immutable environments, because “We can’t trust humans to manipulate production.” So, again, you have the make the path you want them to take, and make this the easiest path. As Eddie notes, “You can’t put too much friction into the pipeline, because developers are good at bypassing friction.” You have to make your CI/CD pipelines the easiest way to get ideas to production by. You also have to abstract complexity and automate policies. You have to have an “opinionated platform, and observe your customer so you can address their pain.”

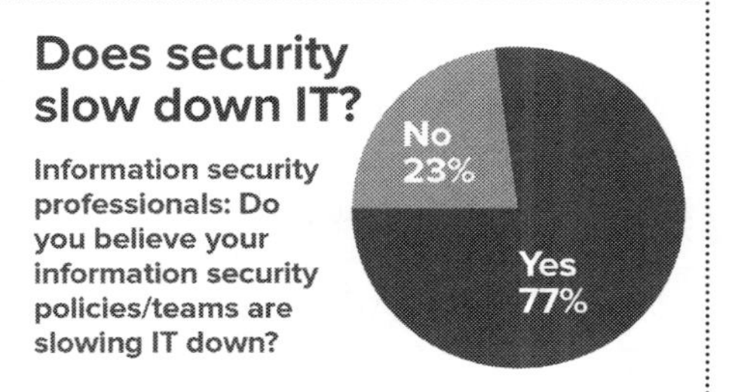

Traditional static application security testing (SAST) and dynamic application security testing (DAST) are too heavyweight, complex and need to be run by a security professional.

This approach won’t work and won’t scale for DevSecOps.

Eddie concluded with the following final points:

- Don’t mandate DevOps. Give employees the chance to master their discipline with examples to set and follow.
- Favor deep end-to-end accomplishments over broad but incremental steps forward. Focus on taking the right teams far before encouraging broad adoption.
- Centralize the platforms and tools that your teams shouldn’t be thinking about. Provide foundational services/commodities and let teams stay on purpose.
- Incorporate contributions from everyone; don’t stifle autonomy. Stay open to new ways of working.
- Challenge security policies, but respect intentions. Find new ways to enforce concerns without abandoning precaution.

Now, go be a disruptor.

CHAPTER 32

Scaling DevOps at Pearson

presented by Sean D. Mack

CHAPTER 32
Scaling DevOps at Pearson

Pearson might be one of the more influential companies you have never heard of. Their footprint in the educational publishing marketplace is expansive, and their scope was just one challenge when they made the move to DevOps.

Sean D. Mack (@SeanDMackNYC) is formerly the VP of Operations and Application for Pearson. To give you an idea of the scale of their operations when they began implementing DevOps, Sean oversaw 110 development teams and 3,000 different products across the globe. To further complicate it, Pearson is a mature company with both legacy and new products. This is a case study of how they approached implementing DevOps with the goal of helping you learn how to do it.

As they say, you often learn your biggest lessons with your failures. Sean noted that, ultimately, DevOps was a failure at Pearson due to organizational change. But, he presented the case study with the hope you will learn from what they did and take it forward. He continues to be a passionate evangelist for DevOps.

Pearson had their challenges. They were a siloed organization operating at a large scale, and they had a toss-it-over-the-wall mentality to dealing with problems. Out of sight, out of mind.

In addition, all of their teams were at different levels of maturity, from legacy teams running .Net on Windows 2000 servers while others were doing CI/CD to AWS. Obviously, this creates challenges when implementing DevOps, but we will see later how they tried to tackle that, specifically.

At the high level, the solution involved several aspects:

- A DevOps Practice Area
- A DevOps Steering Committee

- Measuring progress based on global KPIs focused on measuring business impact
- Measuring teams based on a DevOps scorecard
- Push from the bottom — plenty of developers and operations who embraced and were eager for the change
- Push from the top — support from key executives

Pearson's DevOps Steering Committee set-up several workstreams:

- Tracking tools
- Metrics
- Deployments automation and monitoring practices
- Test automation
- App engineering resource contention
- Operationalization
- Communications

While these are specific to Pearson, many may be applicable to your organization. In any case, it is important to step back, look at your organization, and decide what your organization needs.

Pearson also looked at four core competencies to track and measure: automation; independence — how independent is the feature; operationalization — how easy it is to maintain; and, continuous improvement. The below chart outlines each area in more detail.

Automation	Continuous Improvement	Independence	Operationalization
• Unit tests • Critical Scenarios • Dependent Systems • Continuous Build • Continuous Deployment • Test Automation	• KPI • SLA • Metric Tracking • Metric Reporting • Remediation	• Isolated Configuration • Code Dependencies • Isolated Infrastructure	• Monitoring and Alerting • Tier 1 Service Desk System Knowledge • Tier 2 Service Desk System Knowledge • Documentation

In order to help each team measure their competencies and improvement, Pearson built a self-reporting tool to empower teams but encourage visibility, and they developed business metrics and then a competency model to measure teams against to see how they were progressing. As Sean noted, "We were not trying to get everyone to a 5. Some teams, yes, but we also had legacy products that were running for years, successfully, with little or no maintenance. The investment to bring the teams up wasn't worth the effort. There had to be a business reason."

In the end, Pearson raised DevOps maturity for all of the teams. As noted earlier, though, this was eventually phased out, but Sean presented this case study with the hope you found something to drive DevOps in your company.

In the end, if you have just a couple takeaways, they are this:

- Begin by determining what your desirable outcomes are.
- Don't just do DevOps to be cool. Do it to drive business results.

CHAPTER 33

Securing Immutable Servers in a Serverless World

presented by Erlend Oftedal

CHAPTER 33
Securing Immutable Servers in a Serverless World

Snowflakes are beautiful, unique creations. But, let's keep them in nature. They don't belong in our server infrastructure. Snowflake servers, where every configuration is just a little different, can introduce unnecessary security vulnerabilities and complications. While common in IT infrastructure, in the DevOps realm they are gradually becoming ancient history.

Erlend Oftedal (@webtonull), with Blank and head of the OWASP Norway chapter, discussed the benefits of immutable infrastructure practices within serverless architectures. Erlend walked us through the positive effects on security as well as offered insight on potential problems.

Erlend outlined the progress from physical servers to immutable servers, discussing the benefits of each step and the concerns that lead to the next phase, ending with immutable servers — that is, they can't be changed. Once setup, they are set. Automated processes coupled with immutability principles can eliminate snowflake servers.

Immutability isn't a new concept. Some classic examples include: live Linux CDs, Internet cafes, schools, libraries, hotel business centers, etc. To set up immutable servers, Erlend outlines these steps:

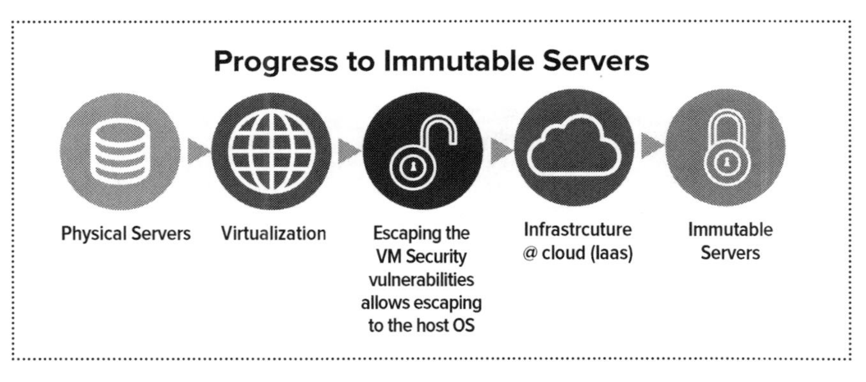

You manage

On premises	IaaS	Containers	FaaS	PaaS	SaaS
Application	Application	Application	Application	Application	Application
Data	Data	Data	Data	Data	Data
Runtime	Runtime	Runtime	Runtime	Runtime	Runtime
OS	OS	OS	OS	OS	OS
Virtualization	Virtualization	Virtualization	Virtualization	Virtualization	Virtualization
Servers	Servers	Servers	Servers	Servers	Servers
Network	Network	Network	Network	Network	Network

You manage ■
Vendor manages □

1. Find a suitable base image
2. Build a specific image for the application
   » Dependencies
   » Application
   » Secrets
3. Deploy image to the server(s)
4. Repeat steps 2-3 for new versions

Of course, this immediately raises questions.

What about data? He notes that because data inherently changes, data has to be stored externally.

How do you deploy changes? Well, you don't change the server. You take one out and put a new one in. What if that one fails? Rollback to the previous one.

How do you manage secrets? You can build secrets in, leverage cloud key management, or use third-party services.

How do we apply security patches? If truly immutable, you deploy a new server with the patch. However, you can have a semi-immutable server that allows for automatically installing patches but still doesn't allow for logins or manual changes.

Speaking of security, what are the security advantages? For starters, if an attacker is in your server, they will be thrown out when you change servers, although they may come back if the vulnerability isn't patched. It also allows for very specialized images by removing all unnecessary packages. With these specialized images, you can more easily audit and monitor the system for unexpected file changes, logins, and connections.

> "Cloud is not JBOS (Just a Bunch of Servers). Amazon is better at running servers than you are."
>
> — DAN KAMINSKY, O'REILLY SECURITY CONFERENCE 2016

Erlend discussed containers because they are a step closer to a serverless world. Containers isolate processes and all run on the same OS. Of course, you can make them immutable, but, whether they are or are not, there are some key security best practices you'll want to consider:

- Don't run as root inside container
- Implement user namespaces so the root account within the container is separated from the root account outside the container
- Utilize the least privileges necessary for capabilities and resources
- Keep them up-to-date
- Automate scanning using tools such as Sonatype Nexus, Twistlock, Clair, or Aqua
- Separate untrusted and trusted containers

Erlend also dove into serverless architecture. While there is debate over what qualifies as serverless, there are 5 principles that Erlend highlighted:

- Use a compute service to execute code on demand
- Write single-purpose, stateless functions
- Design push-based, event-driven pipelines
- Create thicker, more powerful front ends
- Embrace third-party services

Migrating to immutable, serverless infrastructure is a topic to be discussed well beyond this chapter. Whatever you do in the future, though, take his parting words to heart: "the cloud can solve many of your problems, but in the end you cannot transfer your responsibility for security."

CHAPTER 34

System Hardening with Ansible

presented by Akash Mahajan

CHAPTER 34
System Hardening with Ansible

The DevOps pipeline is constantly changing. Therefore relevant security controls must be applied contextually.

We want to be secure, but I think all of us would rather spend our time developing and deploying software. Keeping up with server updates and all of the other security tasks is like cleaning your home — you know it has to be done, but you really just want to enjoy your clean home. The good news is you can hire a "service" to keep your application security up-to-date, giving you more time to develop.

Akash Mahajan (@makash) is a Founder/Director at Appsecco, discussed how to harden your system's security with Ansible. In addition to his role at Appsecco, Akash is also involved as a local leader with the Open Web Application Security Project (OWASP).

Misconfiguration

Akash mentioned the OWASP Top 10 Security Vulnerabilities list, zeroing in on #5 — Security Misconfiguration. To determine if you comply with the guidelines, #5 on the list asks:

- Is any of your software out of date?
- Are there any unnecessary features enabled/installed including ports, services, accounts, pages, or privileges?
- Are default accounts and their passwords enabled/unchanged?
- Are security settings and libraries not set to secure values?

I am sure no one reading this still uses the default administrator password, but can we say the same of your peers? Have you gotten around to installing the latest software patches on your server?

OWASP TOP 10 — A5 Security Misconfiguration

Example Attack Scenarios

- Attacker discovers the standard admin pages are on your server, logs in with default passwords, and takes over.
- Attacker finds due to directory listing and downloads all your compiled Java classes, which she decompiles and reverse engineers to get all your custom code. She then finds a serious access control flaw in your application.
- App server configuration allows stack traces to be returned to users, potentially exposing underlying flaws. Attackers love the extra information error messages provide.

Automation

If a task can be automated, developers automate it. So we should automate our security tasks too, where we can. OWASP provides guidance here, suggesting you should:

- Have a repeatable security hardening process
- Ensure your development, QA, and production servers are configured identically but with different passwords
- Automate the process to minimize the effort required to setup a new secure environment
- Implement a process for deploying all new software updates and patches in a timely manner to each deployed environment
- Run scans and audits periodically to help detect future misconfigurations or missing patches.

This is all part of security hardening, which is, "the process where we identify default configuration present on a system and apply changes that will change the configuration to secure values." This can be applied to your network, transport, application, and kernel networking parameters.

Ansible Playbooks

Ansible is one of the solutions Akash likes to work with, but there are others solutions on the market that provide similar value. Without trying to endorse or evaluate one solution over another, let me share perspectives from Akash's experience with his tool set.

Why does he like it? It boils down to playbooks. An Ansible playbook is a codified security document, allowing you to describe the desired state of a system, rather than the specific steps of how to get to that state. As Akash points out, things change — it is better to have the end state described rather than have to change commands when the system changes.

Other advantages of playbooks include:

- Playbooks are written in YAML providing us with structure that we can learn and train on
- Playbooks are text files, so we can use Git for version control
- Managing playbooks is just like managing any software project
- Playbooks are infrastructure as code but for security
- Playbooks consist of roles, a key aspect of security
- Numerous playbooks are available as open source.

The bottom line is you can, and you should, automate your security hardening process. Your users and other stakeholders will thank you, and, most of all, you will thank yourself because you can spend more time on the things you love to do.

CHAPTER 35

The DevOps Trifecta

presented by Sumit Agarwal

CHAPTER 35
The DevOps Trifecta

DevOps isn't just the right culture, roles, or tools — it is a trifecta. But what is the right combination? Well, that depends on your organization, and it is important to realize that you need all three and have to think strategically about what each one looks like for your organization.

Sumit Agarwal (@aga_sumit) is the Head of Release Engineering at Broadridge Financial Solutions and is leading a change initiative to adopt continuous delivery. His talk strove to help you think about the right culture, roles, and tools for your organization.

Culture

What is DevOps culture? We often hear that DevOps is about *collaboration* and *open communication* between development and operations, and that *is* key. But there is so much more to that...

It is about *sharing responsibility* between development and operations. This means developers are in the on-call rotation, which helps them see how their applications run in production and keeps operations from getting burned out.

It is about *continuous learning*. You run blameless postmortems, because any bug or outage is an opportunity to learn.

It is about *lean practices*. They are pervasive to reduce backslides with more timely feedback and reduce waste by reducing handovers.

And, throughout everything, it is about the eye towards *continuous improvement*: learning from failures, dev understanding ops better and ops understanding dev better, improving communications, etc.

Besides these principles, a DevOps organization must have an effective team and a team that feels safe together (see "Generative" in the following chart).

What Makes an Effective Team

Typology of Organisational Culture – Westrum (2004)

Pathological	Bureaucratic	Generative
Power oriented	Rule oriented	Performance oriented
Low cooperation	Modest cooperation	High cooperation
Messengers shot	Messengers neglected	Messengers trained
Responsibilities shirked	Narrow responsibilities	Risks are shared
Bridging discouraged	Bridging tolerated	Bridging encouraged
Failure → scapegoating	Failure → justice	Failure → inquiry
Novelty crushed	Novelty → problems	Novelty implemented

Roles

With regards to the roles of all team members, you must start with considering operations to be the most important users and a shared responsibility of operational requirements. This allows operations to see how the applications work and to request operational requirements and improvements.

There also needs to be a specific DevOps role. This person orchestrates the DevOps environment, and they should be effective at:

- Coding or scripting
- Process re-engineering
- Communicating and collaborating with others

You have the opportunity to structure a DevOps environment in several ways:

DevOps Silo: This can work for a limited time with the goal to improve communications between Dev and Ops.

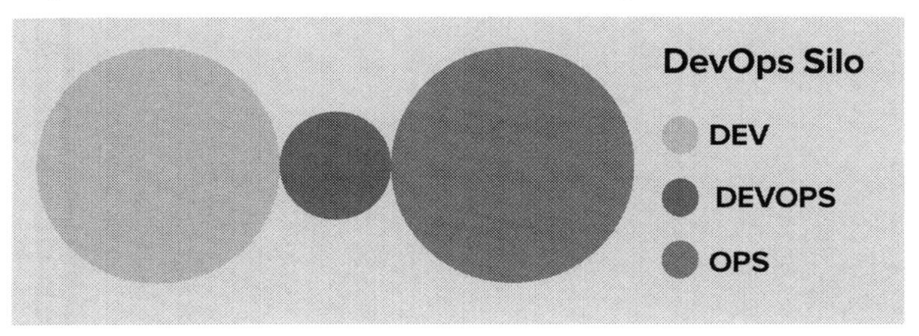

Rebranded Sys Admin: This is effectively just giving a Sys Admin a new title and tools. You should never do this.

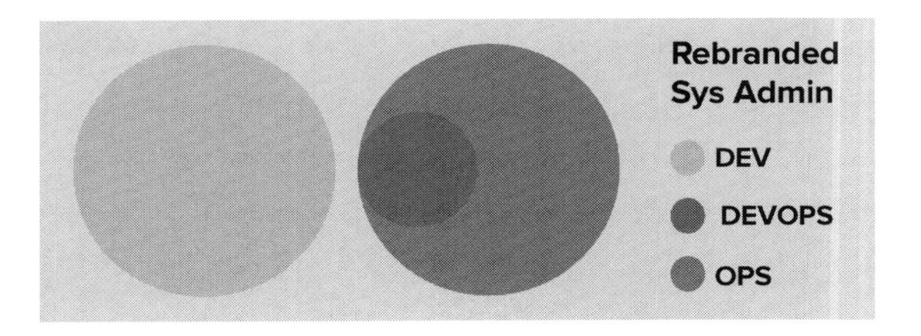

But, what you should really strive for is an
intersection between Dev and Ops:

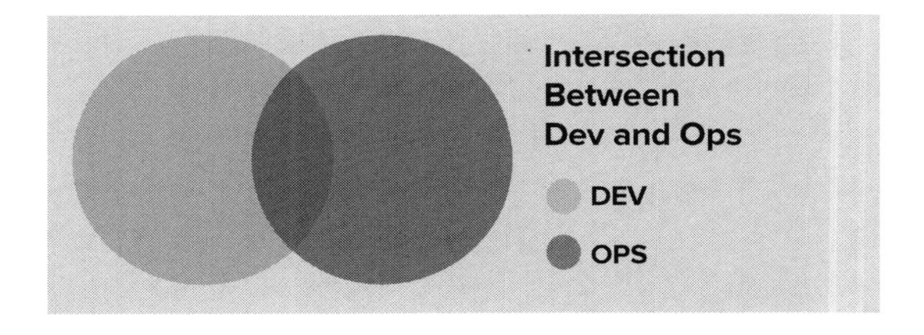

Tools

Remember *Home Improvement* and Tim "The Tool Man" Taylor? He always wanted a bigger and better tool. A power drill with a two-cycle motor? Sure, sounds great. But a "better" tool isn't always better. In fact, Sumit makes the point, "Tools are not essential — look at the problem and see if you are better off building the solution yourself. You will also learn more about the problem and the solutions. Just because you use Jenkins, doesn't mean you are doing continuous integration."

In the end, every DevOps organization needs the right culture, roles, and tools. How do you decide? As Sumit said, "Measure what you want to change and look towards continuous improvement. Retire old metrics that are no longer relevant."

CHAPTER 36

The DevSecOps Equilibrium

presented by Chris Corriere

CHAPTER 36
The DevSecOps Equilibrium

Can you feel the tension in your organization between security, operations, and development? Does each side try to outmaneuver the other? Do they not talk for fear of conflict or being halted in their tracks? You know something needs to be done, but what to do? The answer is simple — everyone needs to be more like pitcher plants. Stay with me.

Chris Corriere is a fascinating guy to talk to over a bowl of ramen or anywhere you might catch him between sessions at a DevOps Days conference. He is one of those guys who search for the deeper meaning of work, relationships, and behaviors. In every conversation, you'll learn something.

In his talk, Chris Corriere (@cacorriere) discusses the Nash equilibrium in relation to security and DevOps environments, shows how nature adapts to similar situations, and presents how we can pull security into a trust relationship, forming DevSecOps.

Every game has a dilemma. Chris explains, "The Sec in DevSecOps means the security folks are explicitly invited to the table. The dilemma is the fact the invitation isn't implied."

In game theory, this fits into the Nash equilibrium — what is commonly illustrated as the Prisoner's Dilemma. You know the setup: two prisoners (A and B) are offered deals to testify against the other, but the deal goes away if prisoner A implicates the B and vice-versa. Although if neither A nor B takes the deal, their sentences will be shorter than if they are both implicated. But, A and B can't talk to each other before deciding.

Chris contends the better illustration is the Stag Hunt. The hunters can work together and potentially get a stag to share for food, but, say one sees a rabbit on the hunt first. They could kill the rabbit and have some guaranteed food, but it would be a much smaller amount and could leave their partner high and dry. Cooperate or compete? Oh, the dilemma!

Chris then presents what he coined the Trinary Nash Equilibria — that each relationship in nature can devolve into: **commensalism**, where one organism benefits but the other one neither benefits or is harmed; **amensalism**, where one organism is inhibited or destroyed while the other is unaffected; or, **parasitism**, where one benefits at the expense of the other. None of these are beneficial for both organisms.

What we want to strive for in our organization is **symbiotism**, a cooperative relationship with high trust and that is beneficial to both parties.

This is seen throughout nature. One example Chris gave comes from low-light, crowded swamps where plants compete for sunlight and nutrients. A species of pitcher plants is shaped so that bats can easily find them with their echolocation cries. The bats roost on the plants, relatively parasite free, and the plant eats their poop. While admittedly gross for you and me, it is a win-win for the bat and the plant.

The DevSecOps lesson for the day: become the pitcher plant — adapt and offer value to unlikely partners.

Of course, human relationships are more complex than pitcher plants and bats. Chris talks for a bit about the Cynefin sense-making Framework by Dave Snowden.

As Chris talked about jungles, ecosystems, and nature, he walked through the value of diversity in nature, making the point that diversity reduces risk, whether in nature or in organizations. Monocultures

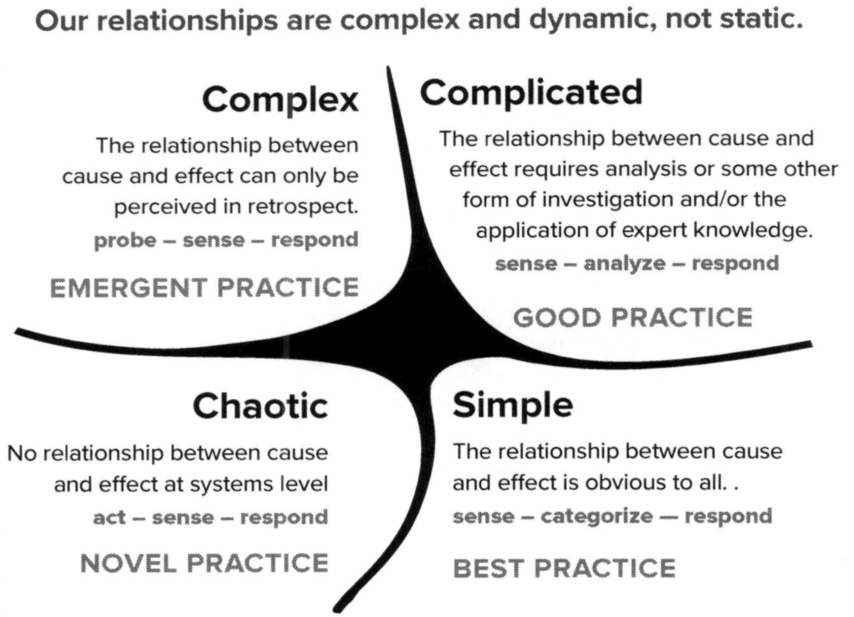

Our relationships are complex and dynamic, not static.

Complex

The relationship between cause and effect can only be perceived in retrospect.

probe – sense – respond

EMERGENT PRACTICE

Complicated

The relationship between cause and effect requires analysis or some other form of investigation and/or the application of expert knowledge.

sense – analyze – respond

GOOD PRACTICE

Chaotic

No relationship between cause and effect at systems level

act – sense – respond

NOVEL PRACTICE

Simple

The relationship between cause and effect is obvious to all. .

sense – categorize — respond

BEST PRACTICE

Cynefin Sense-Making Framework by Dave Snowden

don't survive. In DevSecOps, diversity is more than just combining development, security, and operations. It is about different skill sets, backgrounds, thoughts, beliefs. They combine to make our organizations stronger.

In the end, Chris left us with three takeaways:

- Augment humans with tech instead of replacing them.
- Spend time together. Communicate. Build trust. [hint: this is the most important one]
- Work in diverse teams with mutual goals.

If you happen to be at the same DevOps conference as Chris, seek him out. He has some more interesting illustrations from nature and math to help us better understand and improve our organizations, such as Wardley value chain mapping, replacing Maslow's hierarchy of needs, and Inclusive Collaboration.

CHAPTER 37

The Road to Continuous Deployment

presented by Michiel Rook

CHAPTER 37
The Road to Continuous Deployment

We have all heard the excuses: management won't go for it; the code is a jumbled mess; it's too large; too many regulatory hurdles. The trek to Continuous Integration/Continuous Deployment has stumbled for many enterprises, but many more each day have made it.

Michiel Rook spoke about his road to Continuous Deployment. He used an example he worked on for a large publishing agency in The Netherlands, which operates a number of job portals. This specific project was called the San Diego Project, but it was known internally as the Big Ball of Mud because the code base was such a mess. The image below diagrams the legacy system — the beginning of the road.

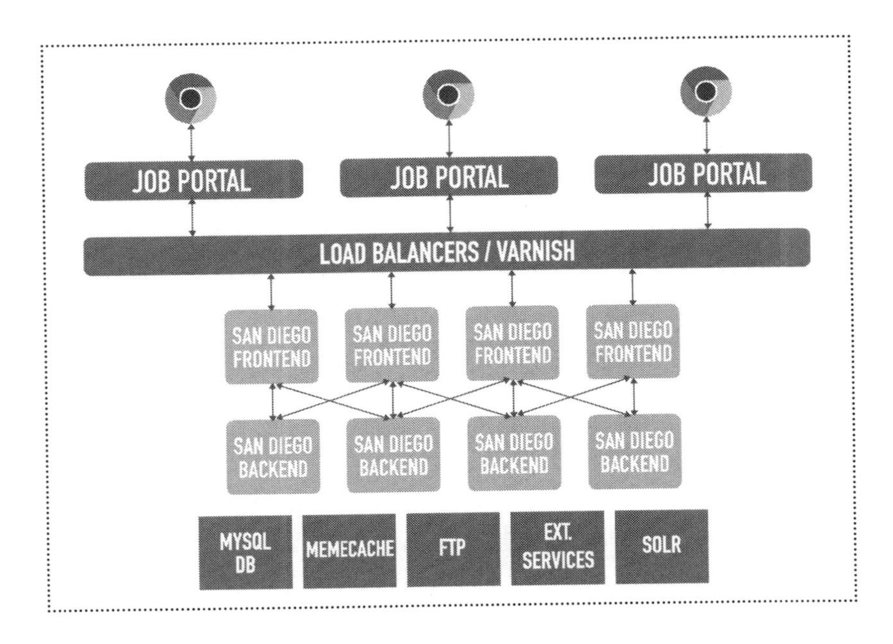

The project was burdened with: infrequent, manual releases, fragile tests, frequent outages and issues, a frustrated team of about 16 people, and low-confidence modifying existing code. Sound familiar? Yep, you were not alone.

The team knew something needed to be done, so they set some goals, including:

- Reduce issues
- Reduce cycle time
- Increase productivity
- Increase motivation

Their approach was to take the monolith, build a proxy and add a service, and then keep adding services until the monolith can be thrown away.

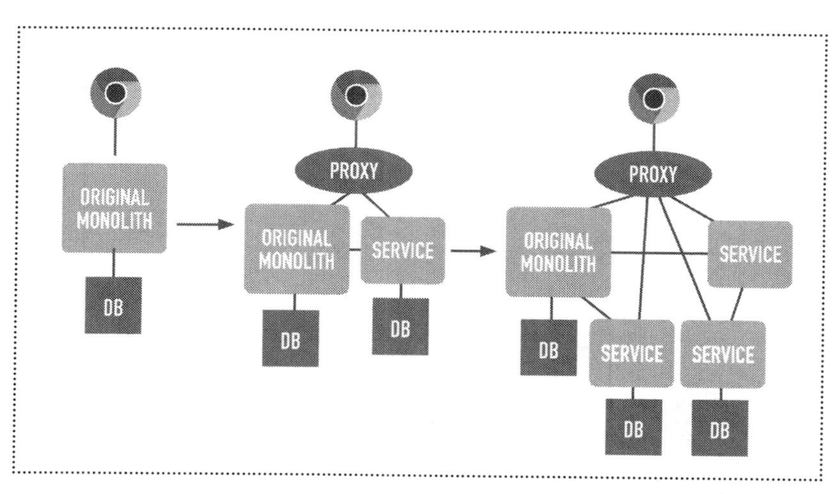

Of course, this is a simple explanation; much more went into the trek.

They started with a foundation of principles to guide their journey:

- Apply the strangler pattern
- Use API first methodology
- Set one service per domain object (job, jobseeker, etc.)
- Migrate individual pages
- Establish services behind load balancers
- Access legacy databases

- Implement continuous deployment
- Utilize Docker containers
- Develop front-ends as services

This all culminates to continuously deliver value, something Michiel calls "Continuous Everything."

It starts with Continuous Integration: developing and building/testing, resulting in an artifact each time. Then, Continuous Delivery: building/testing --> acceptance --> production; going into production is a manual process, but the code is always deployable. Finally, you reach Continuous Deployment when the whole process is automated.

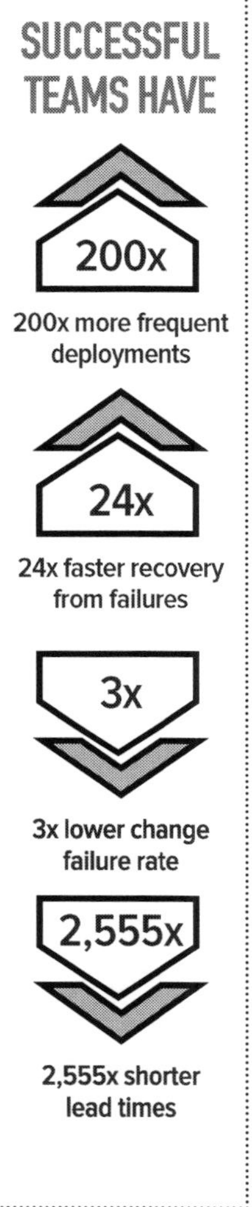

Continuous Deployment is advantageous because it offers the following:

- Small steps
- Early feedback
- Reduce cycle time
- Reduce risk
- Room to experiment

Now that Michiel's project is behind him, he offered the following key aspects of any road trip to continuous deployment:

Only Commit to Master. No branches. You don't want to delay integration and abuse version control for functional separation. Plus, everything on a branch increases the risk of conflicts and delays integration.

Every Commit Goes to Production.

Use Pair Programming for Code Review. This will require discipline, but all development needs to be paired. Mix and match experienced developers.

Quality Gates. Ensure a substantial amount of tests and code coverage.

Feature Toggles and A/B Tests. Determine which version people can/can't see and facilitate A/B testing. But, be sure to keep the number in check.

Dashboards. Display is essential for deployments. Measure everything: KPIs, build times, page load times, number of visitors, results of A/B tests, etc.

DevOps. Mentality is a culture; no more walls between dev and ops. Ownership lies within the team for everything, but this doesn't mean everyone knows everything.

Automate Repeatable Things. If you need to do something twice, you have done it too many times.

Continuous Testing. Use unit tests and smoke tests to see if a service is live, and always monitor. Exploratory testing is important because you continue to test the most critical paths.

Pipeline as Code. Automate the pipeline.

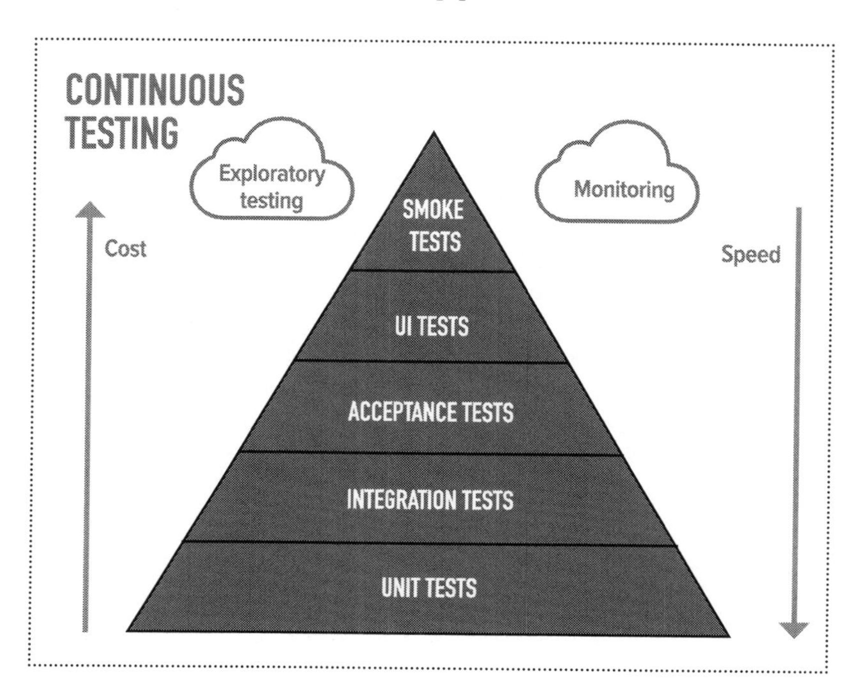

In the end, the deployment looks like this:

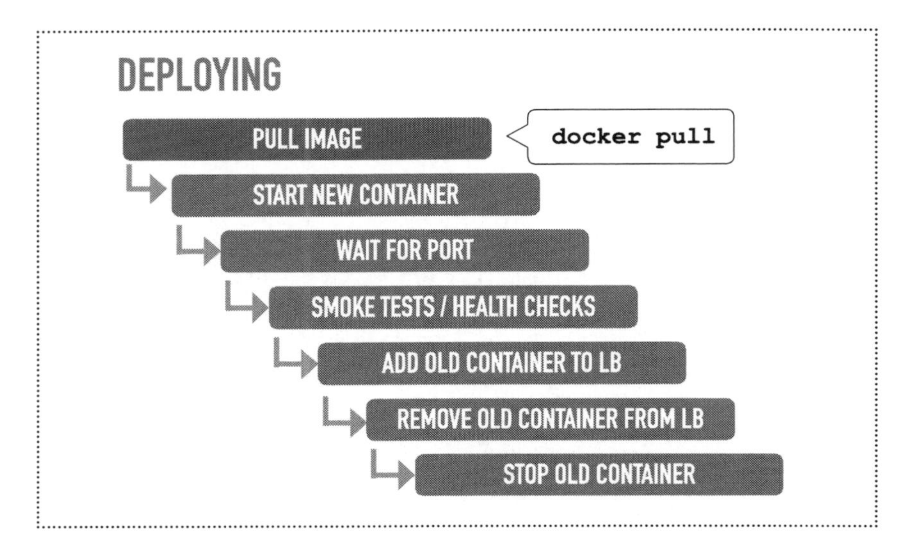

Feedback

DevOps is built on the importance of feedback. One example Michiel had on this project was a large, red flashing light that signaled a build failure. Whenever it went off, that became the number one thing people started working on.

Michiel's project spanned over one year. In the end, they reduced the total build time per service to less than 10 minutes, significantly improved page load times, increased confidence and velocity, and more. The universal truths of the importance of team acceptance and that change is hard were seen. They also learned, among other things, that the alignment with business priorities is key, ensuring you have the necessary experience on staff is essential, and limiting feature toggles is crucial.

Overall, Michiel and his team made it to Continuous Deployment. At the end of his talk, Michiel did report that, to his dismay, the legacy system they are seeking to replace is also still in service. The road is long, but worth the trip.

CHAPTER 38

ABN AMRO Embraced CI/CD to Accelerate Innovation and Improve Security

presented by Stefan Simenon

CHAPTER 38
ABN AMRO Embraced CI/CD to Accelerate Innovation and Improve Security

ABN AMRO is one of the largest banks in the Netherlands. It is a large enterprise that is heavily regulated. They employ 22,000 employees and 5,000 of them work in IT. After a major transformation journey from waterfall to Agile, they now have over 300 Agile teams.

Many organizations would have shied away from the transformation, but ABN AMRO saw FinTech companies nipping at their heels. Transformation was imperative to survival. They couldn't be the technological equivalent of the stereotypical fat cat, cigar smoking, short-work-day bankers who refuse to adapt.

Stefan Simenon was in the middle of the transformation when it began three years ago. His talk focuses on explaining how the bank implemented in CI/CD pipelines to accelerate innovation while maintaining strong governance and security standards.

You don't implement — or even think about implementing — a cultural shift like this in an organization this size because it is a latest trend or you watched some inspiring talks during a recent conference. You have to feel the burden of the status quo. For ABN AMRO, several challenges were staring them in the face:

- Long lead times for software delivery
- Software quality issues found at a late stage
- Many manual handovers and approvals
- Code merges happening late in the dev lifecycle
- Inefficient cooperation between Dev and Ops
- Big non-frequent releases to production

Continuous Integration	Continuous Delivery	Continuous Deployment
Produce automated builds and detect errors as soon as possible, by integrating and testing all changes on a regular (daily) basis.	High frequency delivery of a tested functional piece of software that can be deployed to production rapidly.	Fully automated process including deployment to production without human interaction.

Admitting you have a problem is the first step, but there are many more. As they agreed to move forward with CI/CD, they recognized that CI/CD is about changing the mindset, behaviors, processes, and the "Way of Work" first. The right tool choices would come later.

To proceed, they setup the project organization into a cluster with central and decentralized orientations. The centralized part paved the way by setting up the conditions for the teams to get working. The decentralized parts moved forward by implementing CI/CD within the teams.

Once the teams were in place, they determined they would start with the technologies they had and wait for other tools. They also ensured there was strong alignment between Development, Operations, and Security.

Recognizing that other large organizations often take 3-8 years to implement this level of change and change course along the way, they plan for small milestones at three month intervals while keeping the overall transformation journey in mind. This allowed them to learn and improve as they progress.

One interesting approach they have taken is called "build breakers." That is, once a developer triggers a build and the unit testing is complete, three separate scans are run: a code quality scan with SonarQube; a secure source coding scan with Fortify; and, an open

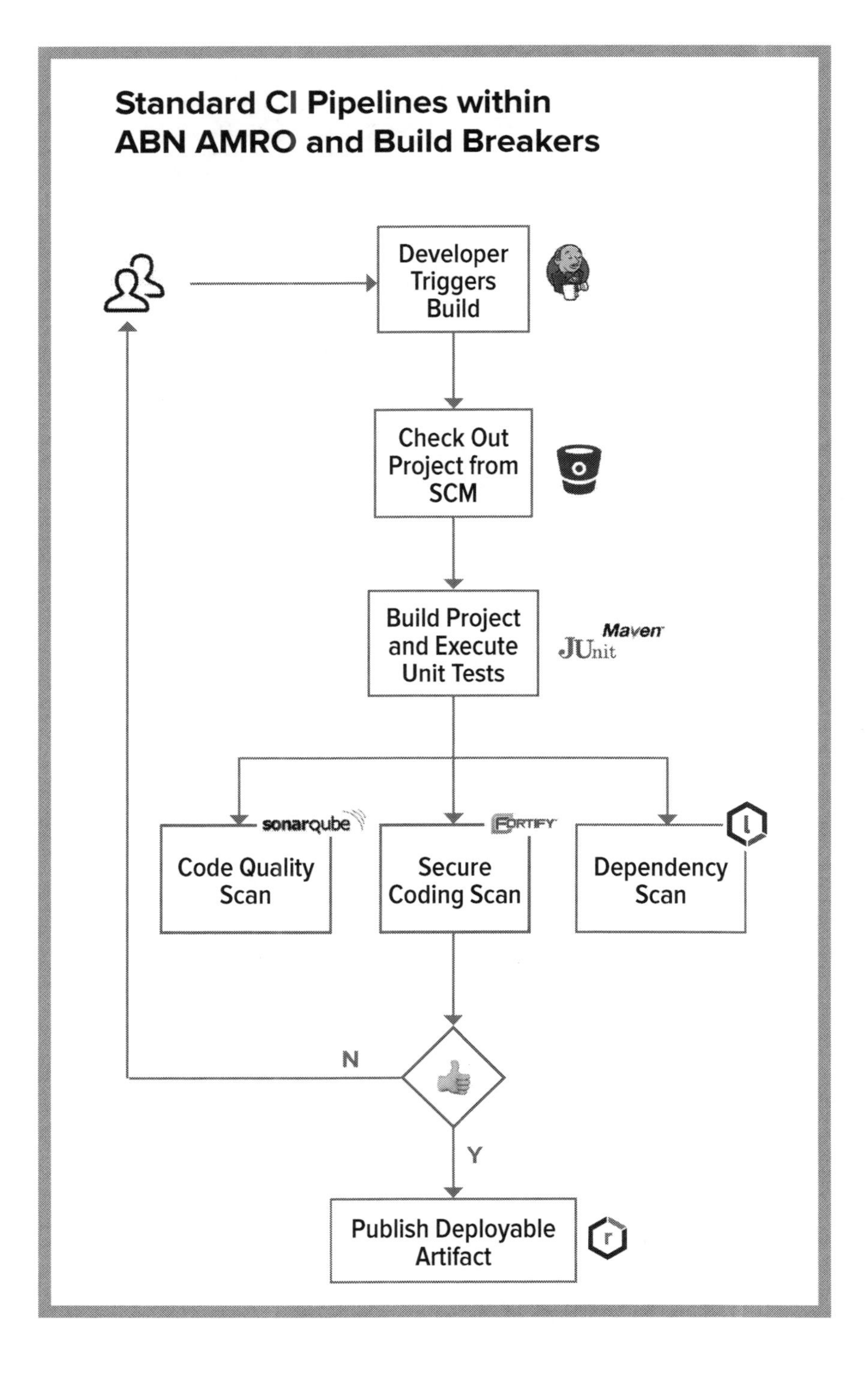
Standard CI Pipelines within ABN AMRO and Build Breakers
Developer Triggers Build
Check Out Project from SCM
Build Project and Execute Unit Tests
Maven
JUnit
sonarqube
FORTIFY
Code Quality Scan
Secure Coding Scan
Dependency Scan
N
Y
Publish Deployable Artifact

source dependency scan with Nexus Lifecycle. A break in any one of these will send the build back to the developer to be fixed.

They also setup an IT for IT organization (IT4IT) to enable CI/CD implementation. The IT4IT organization:

- Implements tooling upgrades
- Implements new tools
- Enhances and improve CI/CD pipelines
- Implements new CI/CD pipelines
- Handles user management
- Supports Agile teams
- Conducts incident and problem management

A lot has happened since they began three years ago. Here are just some of the benefits have they seen so far:

- Test environment uptime improved
- Improved code quality and secure coding
- Improved cooperation across stakeholders
- Improved time to market
- Improved development processes

There is still more to do. As they move forward, they want to further transform to DevOps by improving collaboration between Dev and Ops. They also want to automate and improve tooling pipelines, enhance the IT4IT landscape, implement a hybrid cloud strategy with a mix of internal and AWS clouds, and move toward a service oriented architecture. They also realize that improving the Way of Working, mindsets, and behaviors has to stay top of mind throughout their journey — it is the foundation all of this is built upon.

At the conclusion of his talk, Stefan offered some takeaways:

- Ensure you have senior management and involvement
- Invest in reducing technical debt
- Create a safe environment so people know that failing is okay
- Do not focus just on tooling

- Do not underestimate the journey and complexity
- Do no focus on the long term but rather on small improvements

There is one more takeaway — don't shy away from CI/CD transformation if you are in a large organization. It is possible.

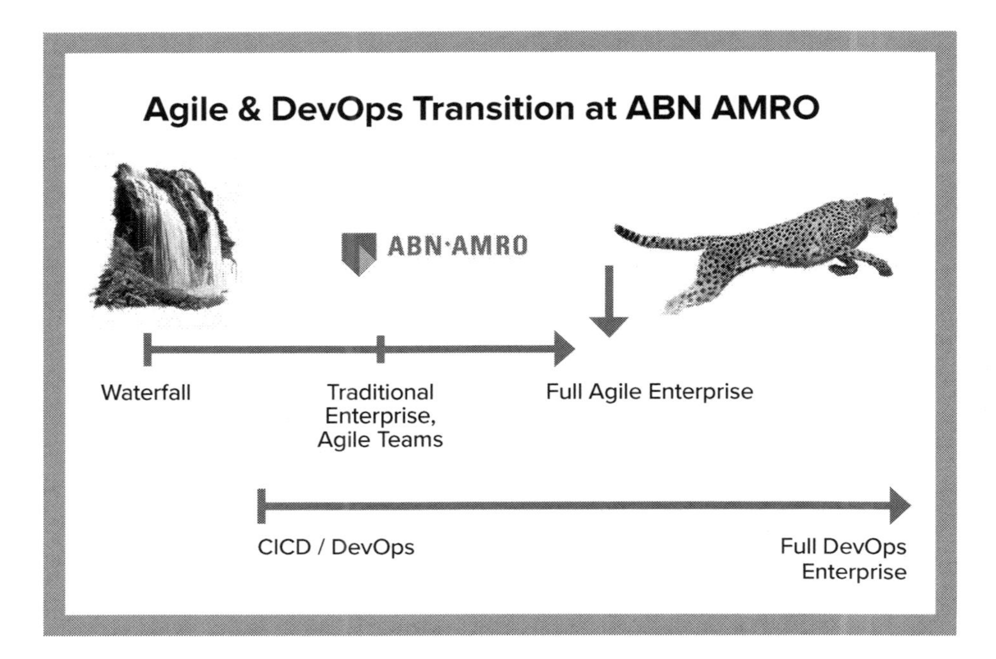

Voices of All Day DevOps, Volume 1

FEEDBACK LOOPS